Lecture Notes in Computer Science 15397

Founding Editors

Gerhard Goos
Juris Hartmanis

The series Lecture Notes in Computer Science (LNCS), including its subseries Lecture Notes in Artificial Intelligence (LNAI) and Lecture Notes in Bioinformatics (LNBI), has established itself as a medium for the publication of new developments in computer science and information technology research, teaching, and education.

LNCS enjoys close cooperation with the computer science R & D community, the series counts many renowned academics among its volume editors and paper authors, and collaborates with prestigious societies. Its mission is to serve this international community by providing an invaluable service, mainly focused on the publication of conference and workshop proceedings and postproceedings. LNCS commenced publication in 1973.

Weizhi Meng · Moti Yung · Jun Shao
Editors

Attacks and Defenses for the Internet-of-Things

7th International Conference, ADIoT 2024
Hangzhou, China, December 13–14, 2024
Proceedings

 Springer

Editors
Weizhi Meng 🆔
Lancaster University
Lancaster, UK

Moti Yung 🆔
Google and Columbia University
New York, NY, USA

Jun Shao 🆔
Zhejiang Gongshang University
Hangzhou, Zhejiang, China

ISSN 0302-9743 ISSN 1611-3349 (electronic)
Lecture Notes in Computer Science
ISBN 978-3-031-85592-4 ISBN 978-3-031-85593-1 (eBook)
https://doi.org/10.1007/978-3-031-85593-1

Preface

The 7th International Conference on Attacks and Defenses for Internet-of-Things (ADIoT 2024) was held on 13–14 December, 2024, in Hangzhou, China.

The Internet of Things (IoT) technology is widely adopted by the vast majority of businesses and is impacting every aspect of the world. However, the natures of the Internet, communication, embedded OS and backend resources make IoT objects vulnerable to cyber attacks. In addition, most standard security solutions designed for enterprise systems are not applicable to IoT devices. As a result, we are facing a big IoT security and protection challenge, and it is urgent to analyze IoT-specific cyber-attacks to design novel and efficient security mechanisms. This conference focused on both sides of IoT attacks and defenses, and sought original submissions that discuss either practical or theoretical solutions to identify IoT vulnerabilities and IoT security mechanisms.

This year, ADIoT received 41 submissions, and each submission was single-blindly reviewed by at least three reviewers. Based on their novelty and quality, 10 regular papers were accepted with an acceptance rate of 24.4%. For the conference program, we had three keynote speakers and two paper sessions. The keynote speakers were: Qiong Huang (South China Agricultural University, China), Guomin Yang (Singapore Management University, Singapore), and Feng Lin (Zhejiang University, China).

For the success of ADIoT 2024, we would like to first thank the authors of all submissions and all the PC members for their great efforts in selecting the papers. We also thank all the external reviewers for assisting the reviewing process.

December 2024

Weizhi Meng
Moti Yung
Jun Shao

Organization

General Co-chairs

Rongxing Lu   University of New Brunswick, Canada
Francesco Palmieri   University of Salerno, Italy
Yizhi Ren   Hangdian University, China

Program Co-chairs

Weizhi Meng   Lancaster University, UK
Moti Yung   Google and Columbia University, USA
Jun Shao   Zhejiang Gongshang University, China

Publicity Chairs

Na Ruan   Shanghai Jiao Tong University, China
Chao Chen   RMIT, Australia

Web Chair

Wei-Yang Chiu   Technical University of Denmark, Denmark

Technical Program Committee

Claudio Ardagna   Università degli Studi di Milano, Italy
Michal Choras   Bydgoszcz University of Science and Technology, Poland
Chao Chen   RMIT, Australia
Elena Doynikova   SPC RAS, Russia
Luca Ferretti   University of Modena and Reggio Emilia, Italy
Jianming Fu   Wuhan University, China
Yunguo Guan   Eastern Michigan University, USA
Yong Guan   Iowa State University, USA
Mehdi Gheisari   Islamic Azad University, Iran

Panayiotis Kotzanikolaou	University of Piraeus, Greece
Dimitris Kavallieros	Centre for Research & Technology Hellas, Greece
Georgios Kambourakis	University of the Aegean, Greece
Wenjuan Li	The Education University of Hong Kong, China
Jay Ligatti	University of South Florida, USA
Yuan Lu	Institute of Software, Chinese Academy of Sciences, China
Xiaobo Ma	Xi'an Jiaotong University, China
Reza Malekian	Malmö University, Sweden
Evgenia Novikova	Saint Petersburg Electrotechnical University, Russia
Josef Pieprzyk	CSIRO/Data61, Australia
Vincenzo Moscato	University of Naples Federico II, Italy
Michele Carminati	Politecnico di Milano, Italy
Kouichi Sakurai	Kyushu University, Japan
Jun Shao	Zhejiang Gongshang University, China
Meng Shen	Beijing Institute of Technology, China
Qingni Shen	Peking University, China
Lei Wang	Shanghai Jiao Tong University, China
Bin Xiao	Hong Kong Polytechnic University, China
Maochao Xu	Illinois State University, USA
Toshihiro Yamauchi	Okayama University, Japan
Guanhua Yan	Binghamton University, USA
Xuyun Zhang	Macquarie University, Australia
Cliff Zhou	University of Central Florida, USA
Cong Zuo	Beijing Institute of Technology, China

Steering Committee

Steven Furnell	University of Nottingham, UK
Sokratis Katsikas	Norwegian University of Science and Technology, Norway
Weizhi Meng (Chair)	Lancaster University, UK

Subreviewers

Nicola Bena	Xu Yang
Filippo Berto	Xin Zhang
Bowen Cui	Fei Zhu
Lin Li	

Contents

An Efficient Edge-Based Privacy-Preserving Range Aggregation Scheme for Aging in Place System

Zhuliang Jia[1], Jinkun Gui[1], Rongxing Lu[1]($\boxtimes$), and Mohammad Mamun[2]

[1] Faculty of Computer Science, University of New Brunswick,
New Brunswick, NB E3B 5A3, Canada
rlu1@unb.ca

[2] National Research Council of Canada, Fredericton, NB E3B 9W4, Canada

Abstract. Aging in place (AiP), as a lifestyle choice policy, has been adopted internationally as a response to population aging. Range aggregation, which is a process that computes the total number of counts for a configured range, is one of the most fundamental methods in AiP to enable the healthcare center to have a comprehensive view of health trends and better monitor the overall health and well-being of the elderly in a given area. However, this aggregation process inevitably introduces security and privacy risks attracting significant research attention. While many privacy-preserving schemes supporting aggregation have been proposed, they either fail to match the range aggregation in the AiP scenario, or incur high computational costs when aggregating numerous data due to the employed homomorphic encryption. To address these challenges, we propose an efficient edge-based privacy-preserving range aggregation scheme for the AiP system. Our scheme employs the superincreasing sequence technique to encode the query vector so that the query user can obtain multiple types of aggregation results within a query and utilizes the one-time matrix encryption technique and the additive secret sharing technique to safeguard sensitive information. Security analysis demonstrates that our proposed scheme can achieve privacy preservation in the range aggregation. In addition, extensive experiments also indicate its high efficiency.

Keywords: Edge-based · Range aggregation · Aging in place · Privacy preservation

1 Introduction

Aging in Place (AiP) is a popular term in social policy, referring to an approach that helps seniors remain in their homes for as long as possible [1]. Extensive literature supports these policies, demonstrating that seniors prefer to stay in their homes as they age [2]. This preference reflects their strong sense of attachment to their home and neighborhood, which contributes to their overall well-being

W. Meng et al. (Eds.): ADIoT 2024, LNCS 15397, pp. 1–21, 2025.
https://doi.org/10.1007/978-3-031-85593-1_1

and social connectedness [3]. Additionally, AiP policies are often motivated by concerns over the high costs associated with residential and nursing homes. By enabling seniors to age in place, these policies aim to provide a more cost-effective and emotionally satisfying solution for elder care. Usually, many Internet of Things (IoT) devices are involved in an AiP scenario, such as wearable sensor devices that transmit health information to nearby smart collector devices or servers. Deploying IoT in AiP settings improves the quality of life for the elderly [4] and offers various benefits, including remote monitoring, assistance for people with reduced mobility, long-term care, enhanced quality of care, and monitoring of vital signs [5]. These technologies enable healthcare to remotely collect and monitor users' health data and improve the quality of elderly life.

The results of range aggregation are crucial for healthcare centers to better monitor the overall health and well-being of the elderly in a given area. Aggregated data within a specific range provides a comprehensive view of health trends within the elderly population, such as computing the total number of patients aged 60 to 80 years old who have chronic diseases. This includes tracking the prevalence of chronic conditions, medication usage, hospital admissions, and other health indicators. Healthcare centers can use this information to identify common health issues and trends. For example, suppose data indicates a high incidence of diabetes across different age ranges in a specific area. In that case, the healthcare center can focus on providing specialized care and preventive measures for diabetes management to elderly individuals within these age ranges. However, this aggregation process introduces security and privacy risks. Data transmitted on third-party networks can lead to data breaches, allowing unauthorized access to sensitive health information and compromising its confidentiality. Additionally, the extensive data collected for aggregation poses privacy concerns, including the potential for invasion of privacy and unauthorized profiling of individuals' health conditions and daily routines. Furthermore, in the AiP scenario, most sensors are resource-constrained but these data need to be immediately collected and processed in time so that the healthcare center can intervene in emergency cases and remote monitor in real-time. As a result, privacy-preserving range aggregation should avoid high computational and communicational costs to maintain efficiency.

To support efficient and privacy-preserving range aggregation queries, numerous schemes have been proposed [6–15]. Table 1 provides a summary of current works, which focus on data aggregation in healthcare. However, these schemes have either limited privacy protection or high computational costs, which makes them unsuitable for practical application in AiP scenarios.

To address the limited computational and privacy performance of existing schemes, we incorporate the assistance of edge devices in our proposed scheme. Furthermore, we adopt the one-time matrix encryption technique to develop a practical range aggregation query scheme that provides privacy-preserving features. Our contributions are summarized as follows:

First, we propose a privacy-preserving range aggregation scheme for the AiP scenario. Our scheme utilizes one-time matrix encryption and additive secret

Table 1. Comparison among existing related schemes

	Privacy Preserving	Range Aggregation	Confidentiality	Edge Assistance	Not Authority	Computational Cost
Han et al. [6]	✓	✓	✓	✗	✗	High
Ren et al. [7]	✓	✓	✓	✗	✗	High
Lu et al. [8]	✓	✓	✓	✗	✗	Medium
Tang et al. [9]	✓	✓	✓	✓	✓	High
Guo et al. [10]	✓	✓	✓	✓	✓	High
Almalki et al. [11]	✓	✓	✓	✓	✓	High
Chakraborty et al. [12]	✓	✓	✓	✓	✓	High
Watkins et al. [13]	✓	✗	✗	✗	✓	Low
Bhowmik et al. [14]	✗	✓	✗	✗	✓	Low
Li et al. [15]	✓	✗	✗	✗	✓	Medium
Our Scheme	✓	✓	✓	✓	✓	Low

sharing techniques to safeguard sensitive information during data aggregation and transmission.

Second, we deploy the idea of superincreasing sequence to design an encoding method, which supports multiple types of aggregation results in a query. By using the superincreasing sequence technique to encode the query vector, the healthcare center can obtain three types of range aggregation results from a single query, namely, count, sum, and average.

Third, we analyze the security of our proposed scheme. The result shows that our proposed scheme is secure under our security model. In addition, we conduct extensive experiments to demonstrate that our scheme operates effectively with numerous IoT devices.

The remainder of this paper is organized as follows. We first illustrate our system model, security model and design goals in Sect. 2, and introduce preliminaries used in our scheme in Sect. 3. In Sect. 4, we describe our scheme. After that, we analyze the security of our scheme in Sect. 5, and in Sect. 6, we evaluate the performance of our proposed scheme. Finally, we draw our conclusions in Sect. 7.

2 Models and Design Goal

In this section, we formalize the system model and security model for our range aggregation scheme and identify our design goal.

2.1 System Model

In our system model, we consider range aggregation queries under a typical edge-based AiP architecture, which mainly involves three types of entities: a set of users $\mathcal{U} = \{U_1, U_2, \cdots, U_n\}$, the edge devices $\mathcal{ED} = \{ED_0, ED_1\}$, and a healthcare center $\mathcal{HC}$, as illustrated in Fig. 1.

- **Users with IoT Devices:** We consider a set of users $\mathcal{U} = \{U_1, \cdots, U_n\}$ registered in our system, and each user $U_i \in \mathcal{U}(i \in [1, n])$ periodically reports their health data $x_i \in Z_m = \{0, \cdots, m-1\}$ to edge devices $\mathcal{ED}$ by using their

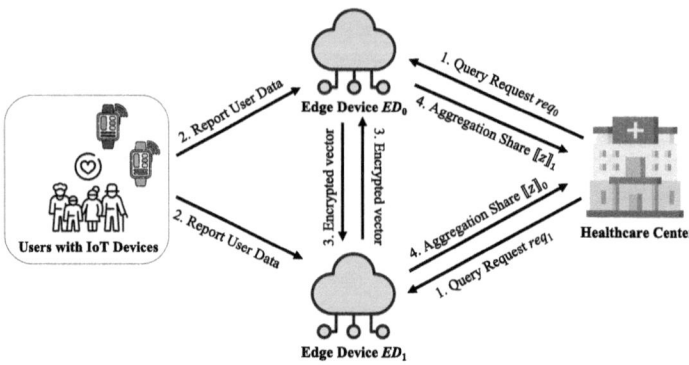

Fig. 1. System model of our scheme

IoT devices (e.g., wearable devices or other resource-constrained sensors). At the same time, since the users' data may contain some sensitive information and edge devices $\mathcal{ED}$ are not fully trustable, users prefer to preserve their privacy data before sending them to $\mathcal{ED}$.

- **Edge Devices** $\mathcal{ED}$: Edge devices $\mathcal{ED} = \{ED_0, ED_1\}$ are situated on the edge of the network, and $\mathcal{ED}$ can be regarded as a link between users $\mathcal{U}$ and the external network in our model. On one hand, they are responsible for processing users' data sent by their IoT devices. On the other hand, they offer the range aggregation query results to the healthcare center.
- **Healthcare Center** $\mathcal{HC}$: The Healthcare Center $\mathcal{HC}$ aims to observe users' health status. To achieve this, the $\mathcal{HC}$ queries the aggregation result within the range $[\alpha, \beta]$, where $0 \leqslant \alpha < \beta \leqslant m - 1$. Three types of range aggregation queries are considered: (1) range count aggregation $Count_{[\alpha,\beta]} = |A|$, (2) range sum aggregation $Sum_{[\alpha,\beta]} = \sum_{x_i \in A} x_i$, and (3) range average aggregation $Average_{[\alpha,\beta]} = \frac{Sum_{[\alpha,\beta]}}{Count_{[\alpha,\beta]}}$, where $A = \{x_i|$ user U_i's data $x_i \in [\alpha, \beta]\}$, $i \in [1, n]$.

2.2   Security Model

In our security model, we consider all users $\mathcal{U}$ to be honest, i.e., they faithfully follow the protocol and send data to edge devices $\mathcal{ED} = \{ED_0, ED_1\}$. Similarly, the healthcare center $\mathcal{HC}$ is viewed as honest, meaning that it will precisely generate query requests. However, edge devices $\mathcal{ED} = \{ED_0, ED_1\}$ are considered to be honest-but-curious, they are interested in both users' private data and the healthcare $\mathcal{HC}$'s queries. One edge device cannot deploy the data-sharing technique, and three or more edge devices will increase additional costs. Therefore, we use two edge devices to balance security and avoid additional overhead. In addition, we consider that there is no collusion between two edge devices ED_0 and ED_1, which is reasonable in the AiP scenario because the risks and costs of being discovered are far outweighed by the potential gains from collusion. Note that, as our scheme focuses on privacy in range aggregation queries, other active attacks, e.g., Denial of Service (DoS) attacks, are beyond the scope of this paper and will be explored in our future work.

2.3 Design Goal

The goal of this paper is to present an efficient and privacy-preserving aggregation range query scheme, which should achieve the following objectives.

- **Privacy Preservation**. Our proposed scheme should 1) preserve each user U_i's private information x_i against the two edge devices $\mathcal{ED}$ and the healthcare center $\mathcal{HC}$; and 2) preserve the healthcare center $\mathcal{HC}$'s aggregation types (i.e., count, sum, average) and specific query v with range $[\alpha, \beta]$ against two edge devices $\mathcal{ED}$ and users.
- **Efficiency**. Since this paper focuses on the time-sensitive Aging in Place scenarios with users with IoT devices and the healthcare center's high real-time requirement, our proposed scheme should be efficient in terms of both computational costs and communication overhead while achieving the above privacy goal.

3 Preliminary

In this section, we introduce some techniques that will be applied in our proposed scheme: Superincreasing Sequence, Additive Secret Sharing, and One-time Matrix Encryption.

3.1 Superincreasing Sequence (SS)

The total number of users is $n \in Z_p^*$, where p is a large prime. The value of each user's data x_i is in the domain of Z_m, where $m \in Z_p$. To enable getting multiple aggregation results in a query, the query user can generate a superincreasing sequence $\boldsymbol{a}$ of size $\kappa \in Z_m$ to construct the query vector $\boldsymbol{Q}$. The generation of a superincreasing sequence is described as follows.

- SsGen(m, n, κ): The generation algorithm takes parameters (m, n, κ) as input, and outputs a sequence $\boldsymbol{a} = (a_1 = 1, a_2, \cdots, a_\kappa)$, where $a_2, \cdots, a_\kappa$ are primes such that

$$\sum_{j=1}^{l-1} a_j \cdot m \cdot n < a_l \tag{1}$$

  for $l = 2, \cdots \kappa$, and $\sum_{j=1}^{\kappa} a_j \cdot m \cdot n \in Z_p$.

3.2   Additive Secret Sharing (ASS)

Additive secret sharing (ASS) [16] is a commonly employed privacy-enhancing technique. The ASS technique can randomly divide a secret $s \in Z_m$ into two shares $\{s_0, s_1\}$ and send them to different users. All the users can retrieve the secret s. Specifically, the ASS technique can be defined as follows.

- AssSha(s): The sharing algorithm is executed by the data owner. It takes the secret s as input and splits it into two random shares $[\![s]\!]_0$ and $[\![s]\!]_1$, then distributes them to different participants U_0, U_1, where $[\![s]\!]_0 + [\![s]\!]_1 \equiv s \bmod m$.
- AssRecon($[\![s]\!]_0, [\![s]\!]_1$): The reconstruction algorithm is executed jointly by the two users with shares. The different users U_0, U_1 input their shares $[\![s]\!]_0, [\![s]\!]_1$ into the reconstruction algorithm, then the algorithm computes $s \equiv [\![s]\!]_0 + [\![s]\!]_1 \bmod m$ and outputs the result s.

3.3   One-Time Matrix Encryption (OME)

One-time matrix encryption (OME) [17] is a common cryptosystem used to encrypt data, with each secret key matrix being valid for a single use, which ensures security under chosen ciphertext attacks. However, if the same secret key matrix is reused to encrypt multiple pieces of data, the OME technique cannot maintain data security under chosen plaintext attacks. For an m-dimensional column vector $\boldsymbol{Q}$, the OME scheme uses an $m \times m$ random matrix to encrypt it, ensuring its confidentiality. It operates through three main processes: key generation, encryption, and decryption. Specifically, the OME technique can be defined as follows.

- OmeSetup(m): The setup algorithm takes the parameter m as input, and outputs a random matrix $\boldsymbol{M}^{-1} \in Z_p^{m \times m}$ as the secret key

$$
\boldsymbol{M}^{-1} = \begin{pmatrix} d_{00} & \cdots & d_{0(m-1)} \\ \vdots & \ddots & \vdots \\ d_{(m-1)0} & \cdots & d_{(m-1)(m-1)} \end{pmatrix}
$$

where p is a large prime, $m \ll Z_p$, and $\boldsymbol{M} \cdot \boldsymbol{M}^{-1} \equiv \boldsymbol{I} \bmod p$, where $\boldsymbol{I}$ represents the identity matrix.
- OmeEnc($\boldsymbol{M}^{-1}, \boldsymbol{Q}$): The encryption algorithm takes the secret key $\boldsymbol{M}^{-1}$ and an m-dimensional column vector $\boldsymbol{Q}$ as input, and outputs a $m \times 1$ ciphertext $CT_{\boldsymbol{Q}} \equiv \boldsymbol{M}^{-1} \cdot \boldsymbol{Q} \bmod p$.
- OmeDec($\boldsymbol{M}, CT_{\boldsymbol{Q}}$): The decryption algorithm takes the ciphertext $CT_{\boldsymbol{Q}}$ and $\boldsymbol{M}$ as input, and outputs the following plaintext $PT_{\boldsymbol{Q}}$:

$$
\begin{aligned} PT_{\boldsymbol{Q}} &= \boldsymbol{M} \cdot CT_{\boldsymbol{Q}} \bmod p \\ &= \boldsymbol{M} \cdot \boldsymbol{M}^{-1} \cdot \boldsymbol{Q} \bmod p \\ &= \boldsymbol{I} \cdot \boldsymbol{Q} \bmod p \\ &= \boldsymbol{Q} \bmod p \end{aligned}
$$

4 Our Proposed Scheme

Our proposed efficient edge-based privacy-preserving range aggregation scheme consists of four phases: 1) query generation; 2) data report; 3) range aggregation; and 4) result reading. We describe each phase in detail.

4.1 Query Generation

In the query generation phase, the healthcare center $\mathcal{HC}$ initiates a query with the range $[\alpha, \beta](0 \leq \alpha < \beta \leq m - 1)$ and sends it to edge devices $\mathcal{ED}$, where $m \in Z_p$. To obtain multiple types of aggregate results in a query, $\mathcal{HC}$ uses the superincreasing sequence technique to encode the query vector. To protect the query range, the $\mathcal{HC}$ uses the OME cryptosystem to safeguard sensitive information. The Algorithm 1 shows the detailed query generation algorithm.

Algorithm 1: Query Generation (QueryGen)

Input: A query with the range (α, β), and the parameter m
Output: Query request req_0, and query request req_1

1 $\mathcal{HC}$:
2 executes to get $\boldsymbol{M_0} \leftarrow \mathrm{OmeSetup}(m)$;
3 executes to get $\boldsymbol{M_1} \leftarrow \mathrm{OmeSetup}(m)$;
4 computes matrix $\boldsymbol{M_0^{-1}}$ such that $\boldsymbol{M_0} \cdot \boldsymbol{M_0^{-1}} = \boldsymbol{I}$;
5 computes matrix $\boldsymbol{M_1^{-1}}$ such that $\boldsymbol{M_1} \cdot \boldsymbol{M_1^{-1}} = \boldsymbol{I}$;
6 runs $\boldsymbol{a} = (a_\alpha, \cdots, a_\beta) \leftarrow \mathrm{SsGen}(m, n, \beta - \alpha + 1)$;
7 initializes $\boldsymbol{Q}^T = (Q_0, Q_2, \cdots, Q_{m-1}) \leftarrow \{0\}^m$;
8 **for** $j \in [0, m - 1]$ **do**
9   **if** $j \in [\alpha, \beta]$ **then**
10         $Q_j = a_j$
11   **else**
12         $Q_j = 0$
13   **end**
14 **end**
15 runs to get $[\![\boldsymbol{Q}]\!]_{M_0^{-1}} \leftarrow \mathrm{OmeEnc}(\boldsymbol{M_0^{-1}}, \boldsymbol{Q})$;
16 runs to get $[\![\boldsymbol{Q}]\!]_{M_1^{-1}} \leftarrow \mathrm{OmeEnc}(\boldsymbol{M_1^{-1}}, \boldsymbol{Q})$;
17 **return** $req_0 = \{[\![\boldsymbol{Q}]\!]_{M_1^{-1}}, \boldsymbol{M_0}\}, req_1 = \{[\![\boldsymbol{Q}]\!]_{M_0^{-1}}, \boldsymbol{M_1}\}$.

Step 1: The $\mathcal{HC}$ inputs the parameter m, running the OME cryptosystem OmeSetup(m) algorithm twice to randomly generate invertible $m \times m$ matrices $\boldsymbol{M_0}$, $\boldsymbol{M_1}$ and computes their inverse $\boldsymbol{M_0^{-1}}$, $\boldsymbol{M_1^{-1}}$.
Step 2: The $\mathcal{HC}$ encodes the range $[\alpha, \beta]$ to a query vector

$$\begin{aligned} \boldsymbol{Q} &= (Q_0, Q_2, \cdots, Q_{m-1})^T \\ &= (0, \cdots, 0, a_\alpha, \cdots, a_\beta, 0, \cdots, 0)^T \end{aligned} \tag{2}$$

using superincreasing sequence technique, i.e., $\mathcal{HC}$ inputs the parameter $(m, n, \beta - \alpha + 1)$, running SS technique SsGen(m, n, κ) to generate a super-increasing sequence $\boldsymbol{a} = (a_\alpha, a_{\alpha+1}, \cdots, a_\beta)$ of length $\kappa = |\boldsymbol{a}| = \beta - \alpha + 1$ which is correspondingly set to α to β dimensions of the query vector $\boldsymbol{Q}$, and the other dimensions of $\boldsymbol{Q}$ are set to 0. These operations are described in lines 6 through 14 in Algorithm 1, and Fig. 2 gives an example of our scheme including how to generate query vector $\boldsymbol{Q}$.

Step 3: After the encoded vectors are generated, the $\mathcal{HC}$ inputs the parameter $(\boldsymbol{M}_0^{-1}, \boldsymbol{Q})$ running OME technique OmeEnc() algorithm to generate encrypted query request $[\![\boldsymbol{Q}]\!]_{\boldsymbol{M}_0^{-1}}$; $\mathcal{HC}$ changes the parameter to $(\boldsymbol{M}_1^{-1}, \boldsymbol{Q})$ to generate encrypted query request $[\![\boldsymbol{Q}]\!]_{\boldsymbol{M}_1^{-1}}$.

Step 4: After the encrypted query vectors are generated, the $\mathcal{HC}$ keeps the inverse matrices $\boldsymbol{M}_0^{-1}$, $\boldsymbol{M}_1^{-1}$ and sends $req_0 = \{[\![\boldsymbol{Q}]\!]_{\boldsymbol{M}_0^{-1}}, \boldsymbol{M}_0\}$, $req_1 = \{[\![\boldsymbol{Q}]\!]_{\boldsymbol{M}_0^{-1}}, \boldsymbol{M}_1\}$ to two edge devices ED_0, ED_1 respectively via secure channels.

4.2   Data Report

In our scheme, each user $U_i \in \mathcal{U} = \{U_1, \cdots, U_n\}$ periodically uses their IoT device to report their data $x_i \in Z_m$ to edge devices $\mathcal{ED}$. To protect users' data, U_i uses the ASS technique to split x_i into random shares $\{[\![\boldsymbol{v}_i]\!]_0, [\![\boldsymbol{v}_i]\!]_1\}$. The detailed process follows these steps:

Algorithm 2: Data Report(DataReport)

Input: $x_1, \cdots, x_n$, parameter m
Output: $[\![\boldsymbol{v}_1]\!]_k, \cdots, [\![\boldsymbol{v}_n]\!]_k$

1 **foreach** *user U_i in $\mathcal{U}$* **do**
2   init $\boldsymbol{v}_i = (v_{i0}, \cdots, v_{i(m-1)}) \leftarrow \{0\}^m$;
3   init $[\![\boldsymbol{v}_i]\!]_0 = ([\![v_{i0}]\!]_0, \cdots, [\![v_{i(m-1)}]\!]_0) \leftarrow \{0\}^m$;
4   init $[\![\boldsymbol{v}_i]\!]_1 = ([\![v_{i0}]\!]_1, \cdots, [\![v_{i(m-1)}]\!]_1) \leftarrow \{0\}^m$;
5   **for** $j \in [0, m-1]$ **do**
6   **if** $j = x_i$ **then**
7               $v_{ij} = 1$
8   **end**
9           $\{[\![\boldsymbol{v}_i]\!]_0, [\![\boldsymbol{v}_i]\!]_1\} \leftarrow$ AssSha(v_{ij}) ;
10   **end**
11   **if** $k = 0$ **then**
12   return $[\![\boldsymbol{v}_i]\!]_0$;
13   **end**
14   **if** $k = 1$ **then**
15   return $[\![\boldsymbol{v}_i]\!]_1$;
16   **end**
17 **end**

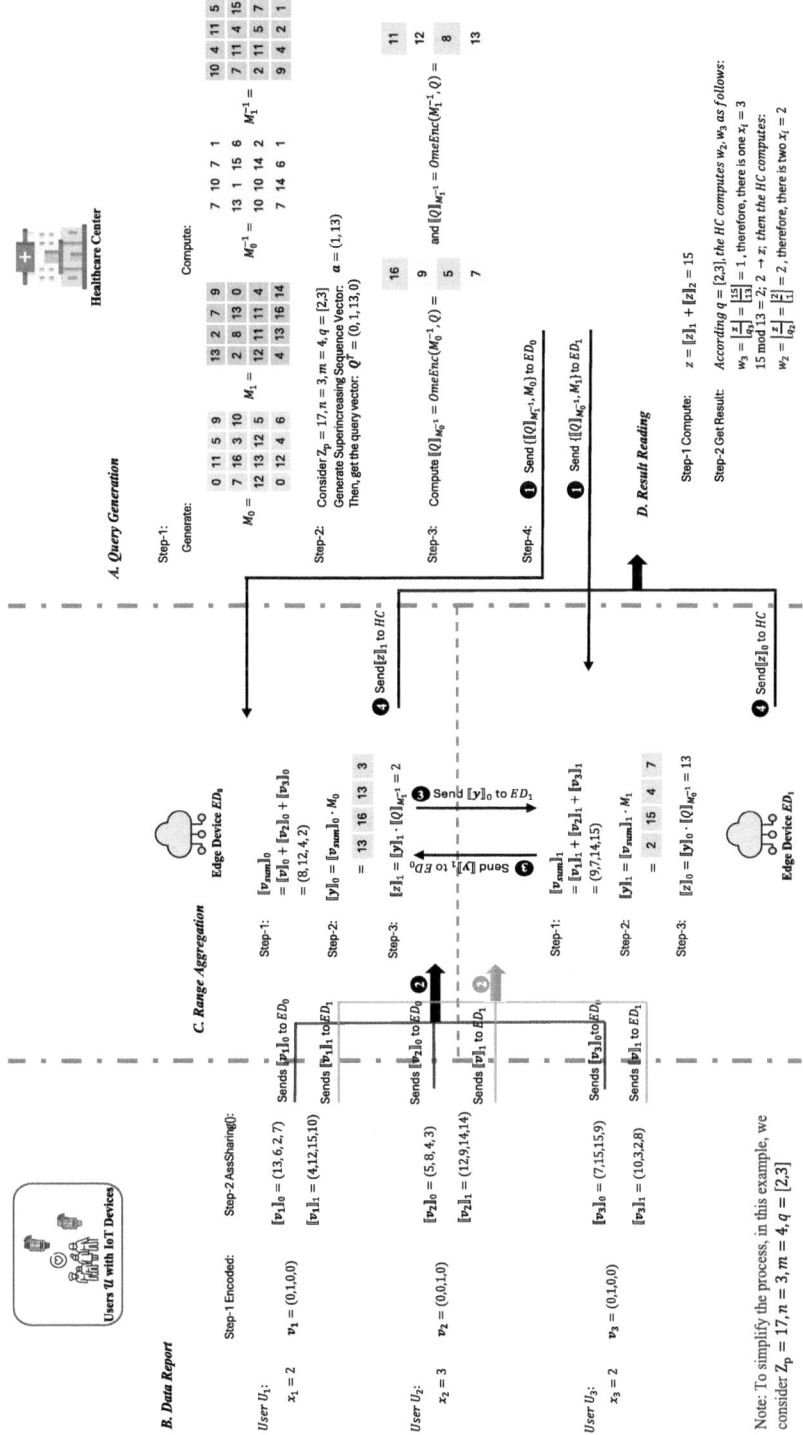

Fig. 2. A simple numeric example of our proposed scheme

Step 1 : User U_i uses the following method encoding x_i to a data vector $v_i = (v_{i0}, \cdots, v_{i(m-1)})$:

$$v_{ij} = \begin{cases} 1, & j = x_i \\ 0, & j \neq x_i \end{cases}, \tag{3}$$

where $j = 0, 1, \cdots, m - 1$.

Step 2 : After generating $v_i = (v_{i0}, \cdots, v_{i(m-1)})$, user $U_i \in \mathcal{U}$ executes ASS AssSha(v_{ij}) algorithm to generate v_i's two random shares vectors $[\![v_i]\!]_0 = ([\![v_{i0}]\!]_0, \cdots, [\![v_{i(m-1)}]\!]_0)$, $[\![v_i]\!]_1 = ([\![v_{i0}]\!]_1, \cdots, [\![v_{i(m-1)}]\!]_1)$ and sends them to ED_0, ED_1 respectively, where $j \in Z_m$.

Correctness. Algorithm 2 reports users' data correctly if and only if (I) Eq. (3) represents data correctly, and (II) for any data $x_i \in Z_m$, the encoded vector $v_i = (v_{i,0}, \cdots, v_{i,m-1})$ by Eq. (3) is unique.

Proof. In the following, we demonstrate the correctness of (I) and (II) successively.

(I) For each user $U_i \in \mathcal{U}$'s data $x_i \in Z_m$, by executing the lines 5 to 10 in Algorithm 2, we get a corresponding data x_i' vector $v_i = (v_{i0}, \cdots, v_{i(m-1)})$, in which the component with 1 represent that x_i equals its index order. Thus, Eq. (3) represents data correctly.

(II) We prove (II) by contradiction. First, suppose that there are two encoded vectors v_1 and v_2 for a user's data x_i.

For vector v_1, after the encoding of Eq. (3), we have that the ηth-dimensional component of v_1 is $v_{1,\eta} = 1$ and other components are 0, where $\eta = x_i$.

For vector v_2, after the encoding of Eq. (3), we have that the δth-dimensional component of v_2 is $v_{1,\delta} = 1$ and other components are 0, where $\delta = x_i$.

Because $\eta = x_i$, $\delta = x_i$, then $\delta = \eta$. Thus, we have

$$(v_{1,0} = v_{2,0}), \cdots, (v_{1,m-1} = v_{2,m-1})$$
$$\Leftrightarrow (v_{1,0}, \cdots, v_{1,m-1}) = (v_{2,0}, \cdots, v_{2,m-1})$$
$$\Leftrightarrow v_1 = v_2$$

Therefore, for any data $x_i \in Z_m$, the encoded vector $v_i = (v_{i,0}, \cdots, v_{i,m-1})$ by Eq. (3) is unique. As a result, Algorithm 2 reports users' data correctly.

4.3   Range Aggregation

Upon receiving query requests and data reports, edge devices $\mathcal{ED} = \{ED_0, ED_1\}$ execute the range aggregation process. To compute the aggregation result privately, edge device $ED_k (k \in \{0, 1\})$ uses Algorithm 3 secure compute range aggregation, which works as follows:

Step 1 : Upon ED_k receiving $([\![v_1]\!]_k, \cdots, [\![v_n]\!]_k)$ from all the users $\mathcal{U} = \{U_1, U_2, \cdots, U_n\}$, ED_k computes

$$[\![v_{sum}]\!]_k = \sum_{i=1}^{n} [\![v_i]\!]_k = [\sum_{i=1}^{n} v_{i0}, \cdots, \sum_{i=1}^{n} v_{i(m-1)}]\!]_k \tag{4}$$

to privately aggregate each dimension of all data.

Step 2 :   After obtaining the share vector $[\![\boldsymbol{v}_{sum}]\!]_k$, Edge device ED_k applies invertible matrix $\boldsymbol{M}_k$ to the vector to obtain $[\![\boldsymbol{y}]\!]_k = [\![\boldsymbol{v}_{sum}]\!]_k \cdot \boldsymbol{M}_k$ and sends it to another edge device ED_{1-k}.

Step 3 :   After receiving $[\![\boldsymbol{y}]\!]_{1-k}$, based on the property of OME OmeDec described in Sect. 3, edge device ED_k gets the aggregation share $[\![z]\!]_{1-k}$ by computing the inner product $[\![z]\!]_{1-k} = [\![\boldsymbol{y}]\!]_{1-k} \cdot [\![\boldsymbol{Q}]\!]_{M_{1-k}^{-1}}$, then ED_k returns $[\![z]\!]_{1-k}$ to $\mathcal{HC}$.

Algorithm 3: Secure Range Aggregation(SRanAgg)

Input: $[\![\boldsymbol{Q}]\!]_{M_{1-k}^{-1}}, \boldsymbol{M}_k, [\![\boldsymbol{v}_1]\!]_k, \cdots, [\![\boldsymbol{v}_n]\!]_k$ to ED_k

Output: $[\![z]\!]_{1-k}$

1 **for** *each ED_k in $\mathcal{ED}$* **do**
2   computes $[\![\boldsymbol{v}_{sum}]\!]_k = \sum_{i=1}^{n} [\![\boldsymbol{v}_i]\!]_k$;
3   computes $[\![\boldsymbol{y}]\!]_k = [\![\boldsymbol{v}_{sum}]\!]_k \cdot \boldsymbol{M}_k$;
4   sends $[\![\boldsymbol{y}]\!]_k$ to ED_{1-k} ;
5   // sends to another edge device;
6   gets $[\![\boldsymbol{y}]\!]_{1-k}$ from another edge device ED_{1-k} ;
7   computes $[\![z]\!]_{1-k} = [\![\boldsymbol{y}]\!]_{1-k} \cdot [\![\boldsymbol{Q}]\!]_{M_{1-k}^{-1}}$;
8   return $[\![z]\!]_{1-k}$ to $\mathcal{HC}$;
9 **end**

Correctness. The following demonstrates that Algorithm 3 correctly computes the aggregation result based on the property of one-time matrix encryption. Since the two edge devices ED_0 and ED_1 are equivalent to each other, we only prove the correctness of ED_0 holds.

Proof. In Algorithm 3, ED_1 computes $[\![\boldsymbol{v}_{sum}]\!]_1 = \sum_{i=1}^{n} [\![\boldsymbol{v}_i]\!]_1 = [\![\sum_{i=1}^{n} v_{i0}, \cdots, \sum_{i=1}^{n} v_{i(m-1)}]\!]_1$ and $[\![\boldsymbol{y}]\!]_1 = [\![\boldsymbol{v}_{sum}]\!]_1 \cdot \boldsymbol{M}_1$. Then, the ED_1 sends $[\![\boldsymbol{y}]\!]_1$ to ED_0. Next, the ED_0 computes $[\![z]\!]_1 = [\![\boldsymbol{y}]\!]_1 \cdot [\![\boldsymbol{Q}]\!]_{M_1^{-1}}$.

Besides, in the query generation phase, to preserve query vector $\boldsymbol{Q} = (Q_0, \cdots, Q_{m-1})^T$ privacy against edge device ED_0, the healthcare center runs $[\![\boldsymbol{Q}]\!]_{M_1^{-1}} \leftarrow \mathrm{OmeEnc}(\boldsymbol{M}_1^{-1}, \boldsymbol{Q})$ to get a $m \times 1$ encrypted vector $[\![\boldsymbol{Q}]\!]_{M_1^{-1}} = \boldsymbol{M}_1^{-1} \cdot \boldsymbol{Q}$. Then, we have that

$$[\![z]\!]_1 = [\![\boldsymbol{y}]\!]_1 \cdot [\![\boldsymbol{Q}]\!]_{\boldsymbol{M}_1^{-1}}$$
$$= [\![\boldsymbol{v}_{sum}]\!]_1 \cdot \boldsymbol{M}_1 \cdot \boldsymbol{M}_1^{-1} \cdot \boldsymbol{Q}$$
$$= [\![\boldsymbol{v}_{sum}]\!]_1 \cdot \boldsymbol{I} \cdot \boldsymbol{Q}$$
$$= \left(\sum_{i=1}^{n} [\![v_{i0}]\!]_1, \cdots, \sum_{i=1}^{n} [\![v_{i(m-1)}]\!]_1 \right) \cdot \begin{pmatrix} Q_0 \\ \vdots \\ Q_{(m-1)} \end{pmatrix} \qquad (5)$$
$$= Q_0 \cdot \sum_{i=1}^{n} [\![v_{i0}]\!]_1 + \cdots + Q_{(m-1)} \cdot \sum_{i=1}^{n} [\![v_{i(m-1)}]\!]_1.$$

Moreover, the $\mathrm{OmeEnc}(\boldsymbol{M}_1^{-1}, \boldsymbol{Q})$ algorithm encrypts an $m \times 1$ dimensional vector $\boldsymbol{Q}$ to an $m \times m$ matrix $[\![\boldsymbol{Q}]\!]_{\boldsymbol{M}_1^{-1}}$. However, this process can not cause a loss of precision in the aggregation result, because after the computation $[\![z]\!]_1 = [\![\boldsymbol{y}]\!]_1 \cdot [\![\boldsymbol{Q}]\!]_{\boldsymbol{M}_1^{-1}}$, due to $\boldsymbol{M}_1 \cdot \boldsymbol{M}_1^{-1} = \boldsymbol{I}$, the aggregation result will regain the accurate result.

4.4   Result Reading

In the result reading phase, the healthcare center $\mathcal{HC}$ first reconstructs aggregation shares received from ED_0, ED_1, then computes different types of aggregation results in need. Algorithm 4 shows the detailed result reading algorithm.

Step 1: After receiving shares $[\![z]\!]_1$, $[\![z]\!]_0$ from ED_0, ED_1, the healthcare center $\mathcal{HC}$ first runs the ASS technique $\mathrm{AssRecon}([\![z]\!]_0, [\![z]\!]_1)$ algorithm to reconstruction the result $z = [\![z]\!]_0 + [\![z]\!]_1 \bmod m$.

Step 2: The healthcare center $\mathcal{HC}$ uses the following iterative operations to compute each count number $w_j (j \in [\alpha, \beta])$ in the query range $[\alpha, \beta]$:

The $\mathcal{HC}$ first sets $j = \beta$, computes $w_j = \lfloor \frac{z}{Q_j} \rfloor$, $z \leftarrow z \bmod Q_j$, then sets $j \leftarrow j-1$ and iteratively computes above operations until $j < \alpha$ or $z \bmod Q_j = 0$ stops, where Q_j is the corresponding query vector element mentioned in step 1 of the query generation phase. These detailed operations are described in lines 5 to 7 in Algorithm 4.

Step 3: Further, after getting $\{w_\alpha, \cdots, w_\beta\}$, the healthcare center $\mathcal{HC}$ can obtain different range aggregation results in need, detailed computation process as follows:

(1) To get the range count aggregation result in the range $[\alpha, \beta]$, the $\mathcal{HC}$ computes $Count_{[\alpha,\beta]} = \sum\limits_{j \in [\alpha,\beta]} w_j$;

(2) To get the range sum aggregation result in the range $[\alpha, \beta]$, the $\mathcal{HC}$ computes $Sum_{[\alpha,\beta]} = \sum\limits_{j \in [\alpha,\beta]} w_j \cdot j$;

(3) To get the range average aggregation result in the range $[\alpha, \beta]$, the $\mathcal{HC}$ computes $Average_{[\alpha,\beta]} = \frac{Sum_{[\alpha,\beta]}}{Count_{[\alpha,\beta]}}$.

Correctness. Algorithm 4 outputs the correct aggregation result if and only if $w_l = \sum_{i=1}^{n} v_{il}$, where $l \in [\alpha, \beta]$.

Algorithm 4: Result Reading(ResultRead)

Input: $[\![z]\!]_1, [\![z]\!]_0, \alpha, \beta, \mathbf{Q}$

Output: $Count_{[\alpha,\beta]}, Sum_{[\alpha,\beta]}, Average_{[\alpha,\beta]}$.

1   executes to get $z \leftarrow$ AssRecon($[\![z]\!]_0, [\![z]\!]_1$);

2   sets $W = \{w_\alpha, \cdots, w_\beta\} = \{0\}^{\beta-\alpha+1}$;

3   sets $j = \beta$;

4   **while** $\alpha \leqslant j \leqslant \beta$ **do**

5   **if** $(j < \alpha) \vee (z = 0)$ **then**

6   break ;

7   **else**

8       $w_j = \lfloor \frac{z}{Q_j} \rfloor$;

9   **end**

10    $z \leftarrow z \bmod Q_j$;

11    $j \leftarrow j - 1$;

12   **end**

13   **if** *wants* $Count_{[\alpha,\beta]}$ **then**

14   outputs $Count_{[\alpha,\beta]} = \sum\limits_{j \in [\alpha,\beta]} w_j$,

15   **end**

16   **if** *wants* $Sum_{[\alpha,\beta]}$ **then**

17   outputs $Sum_{[\alpha,\beta]} = \sum\limits_{j \in [\alpha,\beta]} w_j \cdot j$,

18   **end**

19   **if** *wants* $Average_{[\alpha,\beta]}$ **then**

20   outputs $Average_{[\alpha,\beta]} = \frac{Sum_{[\alpha,\beta]}}{Count_{[\alpha,\beta]}}$.

21   **end**

Proof. The $\mathcal{HC}$ runs $z \leftarrow$ AssRecon($[\![z]\!]_0, [\![z]\!]_1$) algorithm to reconstructe $z = [\![z]\!]_0 + [\![z]\!]_1$. From the Algorithm 3 and its correctness analysis, we know $[\![z]\!]_k = Q_0 \cdot \sum_{i=1}^{n} [\![v_{i0}]\!]_k + \cdots + Q_{(m-1)} \cdot \sum_{i=1}^{n} [\![v_{i(m-1)}]\!]_k$, where $k \in [0,1]$. Thus, we have

$$z = Q_\alpha \cdot \sum_{i=1}^{n} v_{i\alpha} + \cdots + Q_\beta \cdot \sum_{i=1}^{n} v_{i\beta} = \sum_{j=\alpha}^{\beta} Q_j \sum_{i=1}^{n} v_{ij} \qquad (6)$$

In the following, we use mathematical induction to prove that after executing lines 3 to 12 in Algorithm 4, $w_j = \sum_{i=1}^{n} v_{ij}$ holds, where $j \in [\alpha, \beta]$.

From Eq. (1) and Eq. (2), we have $\sum_{j=\alpha}^{l-1} Q_j \cdot m \cdot n < Q_l$, for $l \in [\alpha, \beta]$. Since $\sum_{i=1}^{n} v_{ij} < m \cdot n$, we have $\sum_{j=\alpha}^{l-1} Q_j \cdot \sum_{i=1}^{n} v_{ij} < Q_l$, for $l \in [\alpha, \beta]$.

First, when $l = \beta$, we have $\sum_{j=\alpha}^{\beta-1} Q_j \cdot \sum_{i=1}^{n} v_{ij} < Q_\beta$, thus $w_\beta = \lfloor \frac{z}{Q_\beta} \rfloor = \sum_{i=1}^{n} v_{i\beta}$.

Next, we suppose when $\alpha < l = \delta + 1 < \beta$, $w_{\delta+1} = \sum_{i=1}^{n} v_{i,\delta+1}$ holds. Then, we have the latest $z = \sum_{j=\alpha}^{\delta} Q_j \sum_{i=1}^{n} v_{i\delta}$

Then, when $\alpha < l = \delta < \beta$, we have $\sum_{j=\alpha}^{\delta-1} Q_j \cdot \sum_{i=1}^{n} v_{ij} < Q_\delta$, thus $w_\delta = \lfloor \frac{z}{Q_\delta} \rfloor = \sum_{i=1}^{n} v_{i\delta}$.

Thus, $w_l = \sum_{i=1}^{n} v_{il}$, where $l \in [\alpha, \beta]$. As a result, our Algorithm 4 can output the correct aggregation results.

5   Security Analysis

In this section, we analyze our proposed scheme to demonstrate that it satisfies our privacy requirements.

The security of our scheme is based on the security of additive secret sharing (ASS) and one-time matrix encryption (OME) techniques. As we know, the security of ASS has been proved in [16]. The OME technique was first mentioned in [17]. In a query of our proposed scheme, to enable the OME technique to preserve the query information in privacy, we stipulate that each random secret key matrix is used only once and cannot be reused. Reusing the same matrix cannot preserve data privacy, therefore each time the $\mathcal{HC}$ wants to initiate a query, it should generate two new random matrices to encrypt the query vector. Next, we analyze if our scheme is secure.

In our security model, edge devices $\mathcal{ED} = \{ED_0, ED_1\}$ are considered to be honest-but-curious, they follow the protocol as described but keep a record of each message received, their randomness, and their input. From this information, called the view $\mathbf{View}_k^\pi$ of the kth edge device $k \in \{0, 1\}$, they try to compute the additional information beyond the output during the execution of π. Informally, a protocol π is secure if whatever can be computed by a party participating in the protocol can be computed based on its input and output only. This can be proven by showing the existence of a simulator $\mathcal{S}_k$ that given the kth party's input $\mathbf{input}_k$ and the output $\mathbf{output}_k$ produces a simulation of the view that is indistinguishable from the party's view $\mathbf{View}_k^\pi$ during the protocol, This is formalized according to the simulation paradigm [18].

Definition 1. *Protocol π is secure in the semi-honest model, if there exist simulators $\mathcal{S}_0(\mathbf{input}_0, \mathbf{output}_0)$ and $\mathcal{S}_1(\mathbf{input}_1, \mathbf{output}_1)$, such that*
$$\mathbf{View}_0^\pi(\mathbf{input}_0, \mathbf{input}_1) \stackrel{c}{\equiv} \mathcal{S}_0(\mathbf{input}_0, \mathbf{output}_0)$$
$$\mathbf{View}_1^\pi(\mathbf{input}_0, \mathbf{input}_1) \stackrel{c}{\equiv} \mathcal{S}_1(\mathbf{input}_1, \mathbf{output}_1)$$

where $\stackrel{c}{\equiv}$ denotes computational indistinguishability.

We use the simulation paradigm to prove that our scheme is secure for edge devices, that is to prove the security of the SRanAgg protocol. To facilitate understanding, we use the following Theorem 1 to present in detail the analysis process of the security of the SRanAgg protocol.

Theorem 1. *As long as the additive secret sharing (ASS) and one-time matrix encryption (OME) techniques are secure against semi-honest adversaries, our proposed scheme with the leakage $\mathcal{L}$ is secure under the semi-honest model.*

Proof. We first define the leakage function $\mathcal{L} = \{m, n, p\}$, where m is the dimension of the query vector, n is the total number of users, and p is a large prime.

Based on $\mathcal{L}$, we can define the ideal/real world as follows.

We analyze that our scheme is secure for ED_0. In the ideal world, there is a probabilistic polynomial time adversary, denoted as $\mathcal{A}_0$, and a simulator $\mathcal{S}_0$ with the leakage $\mathcal{L}$.

- *Query generation phase*: The simulator $\mathcal{S}_0$ selects a random matrix $M'_0 \in Z_p^{m \times m}$ and an m-dimensional random vector $[\![Q]\!]'_{M_1^{-1}} \in Z_p^m$, then sends M'_0, $[\![Q]\!]'_{M_1^{-1}}$ to $\mathcal{A}_0$.
- *Data report phase*: For each $i \in \{1, n\}$, the simulator $\mathcal{S}_0$ selects random vector $[\![v_i]\!]'_0 = ([\![v_{i0}]\!]'_0, \cdots, [\![v_{i(m-1)}]\!]'_0)$, and sends $[\![v_1]\!]'_0, \cdots, [\![v_n]\!]'_0$ to $\mathcal{A}_0$.
- *Range aggregation phase*: $\mathcal{S}_0$ replaces $[\![y]\!]_1$ with a random number $[\![y]\!]'_1$ selected from the domain of Z_p, and sends $[\![y]\!]'_1$ to $\mathcal{A}_0$. Therefore, in the ideal world, the view of $\mathcal{A}_0$ is that

$$\mathbf{View}_{\mathcal{A}_0, ideal} = \{[\![Q]\!]'_{M_1^{-1}}, M'_0, [\![v_1]\!]'_0, \cdots, [\![v_n]\!]'_0, [\![y]\!]'_1\}.$$

In the real world, the view of adversary $\mathcal{A}_0$ is as follows:

$$\mathbf{View}_{\mathcal{A}_0, real} = \{[\![Q]\!]_{M_1^{-1}}, M_0, [\![v_1]\!]_0, \cdots, [\![v_n]\!]_0, [\![y]\!]_1\}.$$

Based on the security of OME encryption, $\mathcal{A}_0$ cannot distinguish $\{[\![Q]\!]_{M_1^{-1}}, M_0\}$ and $\{[\![Q]\!]'_{M_1^{-1}}, M'_0\}$. Based on the security of ASS encryption, $\mathcal{A}_0$ cannot distinguish $\{[\![v_1]\!]_0, \cdots, [\![v_n]\!]_0, [\![y]\!]_1\}$ and $\{[\![v_1]\!]'_0, \cdots, [\![v_n]\!]'_0, [\![y]\!]'_1\}$. Therefore, $\mathcal{A}_0$ cannot distinguish the view of real and ideal experiments.

Next, we analyze that our scheme is secure for ED_1. In the ideal world, there is a probabilistic polynomial time adversary, denoted as $\mathcal{A}_1$, and a simulator $\mathcal{S}_1$ with the leakage $\mathcal{L}$.

- *Query generation phase*: The simulator $\mathcal{S}_1$ selects a random matrix $M'_1 \in Z_p^{m \times m}$ and an m-dimensional random vector $[\![Q]\!]'_{M_0^{-1}} \in Z_p^m$, then sends M'_1, $[\![Q]\!]'_{M_0^{-1}}$ to $\mathcal{A}_1$.
- *Data report phase*: For each $i \in \{1, n\}$, the simulator $\mathcal{S}_1$ selects random vector $[\![v_i]\!]'_1 = ([\![v_{i0}]\!]'_1, \cdots, [\![v_{i(m-1)}]\!]'_1)$ and sends them to $\mathcal{A}_1$.
- *Range aggregation phase*: $\mathcal{S}_1$ replaces $[\![y]\!]_0$ with a random number $[\![y]\!]'_0$ selected from the domain of Z_p, and sends $[\![y]\!]'_0$ to $\mathcal{A}_1$. Therefore, in the ideal world, the view of $\mathcal{A}_1$ is that

$$\mathbf{View}_{\mathcal{A}_1, ideal} = \{[\![Q]\!]'_{M_0^{-1}}, M'_1, [\![v_1]\!]'_1, \cdots, [\![v_n]\!]'_1, [\![y]\!]'_0\}.$$

In the real world, the view of $\mathcal{A}_1$ is that

$$\mathbf{View}_{\mathcal{A}_1, real} = \{[\![Q]\!]_{M_0^{-1}}, M_1, [\![v_1]\!]_1, \cdots, [\![v_n]\!]_1, [\![y]\!]_0\}.$$

Based on the security of OME encryption, $\mathcal{A}_1$ cannot distinguish $\{[\![Q]\!]_{M_0^{-1}}, M_1\}$ and $\{[\![Q]\!]'_{M_0^{-1}}, M'_1\}$. Based on the security of ASS encryption, $\mathcal{A}_1$ cannot distinguish $\{[\![v_1]\!]_1, \cdots, [\![v_n]\!]_1, [\![y]\!]_0\}$ and $\{[\![v_1]\!]'_1, \cdots, [\![v_n]\!]'_1, [\![y]\!]'_0\}$. Therefore, $\mathcal{A}_1$ cannot distinguish the view of real and ideal experiments. In conclusion, our proposed scheme is secure under our security model.

6   Performance Evaluation

In this section, we experimentally evaluate the performance of our range aggregation scheme in terms of computational costs and communication overhead.

6.1   Experiment Setting

We implement all schemes in Java and execute experiments on an Apple M1 CPU @ 3.2 GHz Mac Platform with 8 GB RAM. As described in our system model, users' data were considered to be integers, therefore users' data were simulated by generating random numbers, specifically in the Java integer data type. Similarly, the query range was considered to be in an integer range, it was simulated by generating random numbers in the Java integer data type. The matrices of the OME technique used to encrypt users' data and query were considered to consist of real numbers, therefore these matrices were instantiated using the Java double data type.

6.2   Computation Cost

In the query generation phase conducted by the $\mathcal{HC}$, the primary computational demand is to run the OME cryptosystem to generate two pairs of $m \times m$ invertible matrices and use these to encrypt m-dimensional query vector $\boldsymbol{Q}$. Then, in the data report phase conducted by the users, the computational cost primarily arises from each user U_i to split data x_i's into $\{[\![\boldsymbol{v}_i]\!]_0, [\![\boldsymbol{v}_i]\!]_1\}$, where $\boldsymbol{v}_i$ is an m-dimensional vector. After that, in the range aggregation phase, on the edge devices $\mathcal{ED}$, the primary computational demand is to execute inner operations based on ciphertext, i.e., $[\![\boldsymbol{y}]\!]_k = [\![\boldsymbol{v}_{sum}]\!]_k \cdot \boldsymbol{M}_k$ and $[\![z]\!]_{1-k} = [\![\boldsymbol{y}]\!]_{1-k} \cdot [\![\boldsymbol{Q}]\!]_{M_{1-k}^{-1}}$. In the final result reading phase, on the $\mathcal{HC}$, the main computation cost is as follows: the first is to compute $z = [\![z]\!]_0 + [\![z]\!]_1 \bmod m$; and then iteratively compute $w_j = \lfloor \frac{z}{Q_j} \rfloor$, $z \leftarrow z \bmod Q_j$ to get different types of aggregation results.

According to the above theoretical analysis, our scheme's computational efficiency is mainly affected by two parameters: the number of devices n and the data range m. Therefore, to evaluate the efficiency of our scheme, we first test the execution time by changing the number of IoT devices n, the test results are shown in Fig. 3, which shows that the execution time on the healthcare center and edge devices increases linearly with the number of IoT devices n, while the execution time on the IoT devices does not vary with the number of IoT devices n. This is because the amount of data received from IoT devices that need to be aggregated is related to n; in the result reading phase, the healthcare center executes mod Q_j operation to read aggregate results, while according to the Superincresing Sequence (SS) technique, n is one of the factors affecting Q_j. Since IoT devices run in parallel, n does not affect the execution time of IoT devices.

Then, we test the execution time by changing the data range m, the test results are shown in Fig. 4, which shows that the execution time on each participant increases linearly with the number of users n. This is because, for the

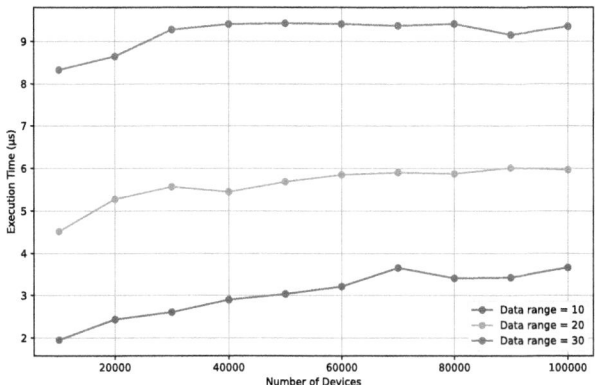

(a) The execution time of the healthcare center

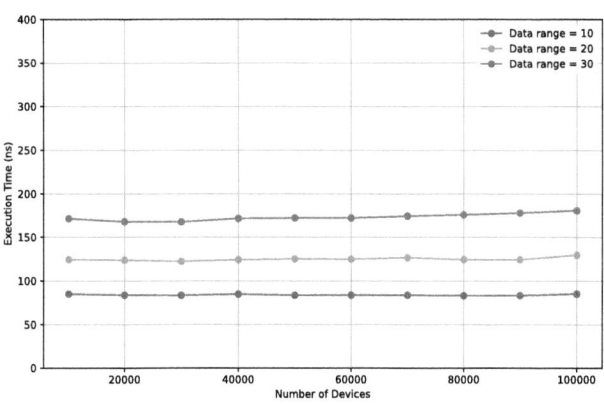

(b) The execution time of users

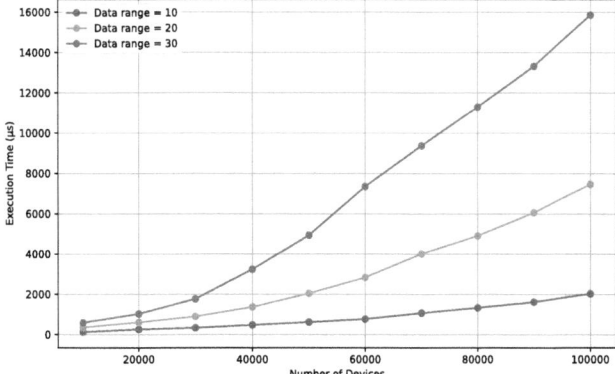

(c) The execution time of edge devices

Fig. 3. The execution time varies with the number of devices

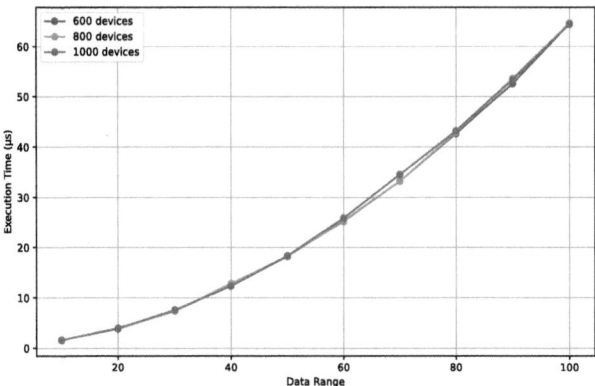

(a) The execution time of the healthcare center

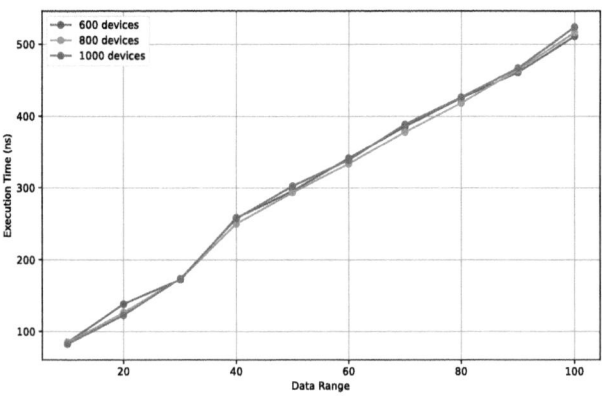

(b) The execution time of users

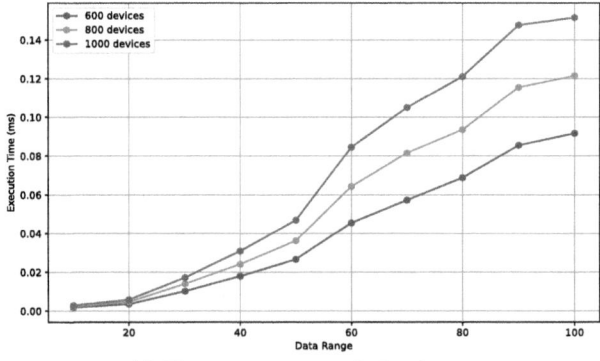

(c) The execution time of edge devices

Fig. 4. The execution time varies with the data range

healthcare center, m affects the dimension of the query vector and OME matrices, and the mod Q_j operation is affected by m; For users, the execution time of splitting data is affected by m; For edge devices, the primary computational overhead, that is execute inner operations based on ciphertext, is affected by m.

6.3   Communication Overhead

For the communication overhead, in the query generation phase, the healthcare center $\mathcal{HC}$ transmits $\{[\![Q]\!]_{M_1^{-1}}, M_0\}, \{[\![Q]\!]_{M_0^{-1}}, M_1\}$ to two edge devices ED_0 and ED_1 respectively, where the Q is m-dimensional query vector and M_0, M_1 is $m \times m$ random invertible matrices. In the data report phase, each IoT device D_i sends v_i's two random shares $\{[\![v_i]\!]_0, [\![v_i]\!]_1\}$ to ED_0, ED_1 respectively, where the shares can be regarded as m-dimensional vectors. In the range aggregation phase, the main communications include each edge device $ED_k (k \in \{0, 1\})$ sends $[\![y]\!]_k$ to ED_{1-k}, gets $[\![y]\!]_{1-k}$ from another edge device, and sends $[\![z]\!]_{1-k}$ to $\mathcal{HC}$, where $[\![y]\!]_k, [\![y]\!]_{1-k}$ can be regard as m-dimensional vectors, $[\![z]\!]_{1-k}$ is an integer. Table 2 presents the communication overhead result of our scheme in each phase.

Table 2. Communication overhead of each phase

Query Generation	Data Report	Range aggregation	Result Reading
$2(m^2 + m)$	$2nm$	$2(2m + 1)$	0

7   Conclusion

In this paper, we have proposed an efficient edge-based privacy-preserving range aggregation scheme for the Aging in Place system. Specifically, we used the superincreasing sequence technique to encode the query vector so that the query user can obtain multiple types of aggregate results in a query. We utilized the additive secret sharing technique and one-time matrix encryption technique to safeguard sensitive information during the range aggregation process. Detailed security analysis confirms that our proposed scheme meets the desirable privacy requirements. Furthermore, our performance evaluation also indicates that our proposed scheme is efficient. In the future, we will explore high dimensions and more different types of range queries on AiP systems.

Acknowledgments. This project was supported by collaborative research funding from the National Research Council of Canada's Aging in Place Challenge Program.

References

1. Lewis, C., Buffel, T.: Aging in place and the places of aging: a longitudinal study. J. Aging Stud. **54**, 100870 (2020). https://doi.org/10.1016/j.jaging.2020.100870
2. Means, R.: Safe as houses? Ageing in place and vulnerable older people in the UK. Soc. Policy Admin. **41**, 65–85 (2007)
3. Wiles, J.L., Leibing, A., Guberman, N., Reeve, J., Allen, R.E.S.: The meaning of "aging in place" to older people. Gerontologist **52**(3), 357–366 (2011)
4. Kumar, R., Jain, A., Tripathi, A.K., Tyagi, S.: COVID-19 outbreak: an epidemic analysis using time series prediction model. In: 2021 11th International Conference on Cloud Computing, Data Science and Engineering (Confluence), pp. 1090–1094 (2021)
5. Pani-Harreman, K.E., Bours, G.J.J.W., Zander, I., Kempen, G.I.J.M., van Duren, J.M.A.: Definitions, key themes and aspects of 'ageing in place': a scoping review. Ageing Soc. **41**(9), 2026–2059 (2021). https://doi.org/10.1017/S0144686X20000094
6. Han, S., Zhao, S., Li, Q., Ju, C., Zhou, W.: PPM-HDA: privacy-preserving and multifunctional health data aggregation with fault tolerance. IEEE Trans. Inf. Forensics Secur. **11**(9), 1940–1955 (2016). https://doi.org/10.1109/TIFS.2015.2472369
7. Ren, H., Li, H., Liang, X., He, S., Dai, Y., Zhao, L.: Privacy-enhanced and multifunctional health data aggregation under differential privacy guarantees. Sensors **16**(9), 1463 (2016). https://doi.org/10.3390/s16091463
8. Lu, R., Heung, K., Habibi Lashkari, A., Ghorbani, A.A.: A lightweight privacy-preserving data aggregation scheme for fog computing-enhanced IoT. IEEE Access **5**, 3302–3312 (2017). https://doi.org/10.1109/ACCESS.2017.2677520
9. Tang, W., Ren, J., Deng, K., Zhang, Y.: Secure data aggregation of lightweight e-healthcare IoT devices with fair incentives. IEEE Internet Things J. **6**(5), 8714–8726 (2019)
10. Guo, C., Tian, P., Choo, K.-K.R.: Enabling privacy-assured fog-based data aggregation in E-healthcare systems. IEEE Trans. Industr. Inf. **17**(3), 1948–1957 (2021). https://doi.org/10.1109/TII.2020.2995228
11. Almalki, F.A., Othman, S.B.: EPPDA: an efficient and privacy-preserving data aggregation scheme with authentication and authorization for IoT-based healthcare applications. Wirel. Commun. Mob. Comput. **2021**, 5594159:1–5594159:18 (2021)
12. Chakraborty, C., Othman, S.B., Almalki, F.A., Sakli, H.: FC-SEEDA: fog computing-based secure and energy efficient data aggregation scheme for Internet of healthcare Things. Neural Comput. Appl. **36**(1), 241–257 (2024)
13. Watkins, M., Dorsey, C., Rennier, D., Polley, T., Sherif, A., Elsersy, M.: Privacy-preserving data aggregation scheme for e-health. In: International Conference on Emerging Technologies and Intelligent Systems, pp. 638–646. Springer, Cham (2022)
14. Bhowmik, T., Mojumder, R., Banerjee, I., Das, G.: IoT based data aggregation method for e-health monitoring system. In: 12th International Conference on Computing Communication and Networking Technologies (ICCCNT), pp. 1–7. IEEE, Kharagpur, India (2021). https://doi.org/10.1109/ICCCNT51525.2021.9579885
15. Li, C.-T., Shih, D.-H., Wang, C.-C., Chen, C.-L., Lee, C.-C.: A blockchain based data aggregation and group authentication scheme for electronic medical system. IEEE Access **8**, 173904–173917 (2020). https://doi.org/10.1109/ACCESS.2020.3025898

16. Blundo, C., De Santis, A., Vaccaro, U.: Efficient sharing of many secrets. In: Enjal-bert, P., Finkel, A., Wagner, K. W. (eds.) STACS 1993. LNCS, vol. 665, pp. 692–703. Springer, Heidelberg (1993). https://doi.org/10.1007/3-540-56503-5_68
17. Hill, L.S.: Cryptography in an algebraic alphabet. Am. Math. Mon. **36**(6), 306–312 (1929)
18. Lindell, Y., Pinkas, B.: A proof of security of Yao's protocol for two-party com-putation. J. Cryptol. **22**(2), 161–188 (2009). https://doi.org/10.1007/S00145-008-9036-8

An Empirical DNN Pruning Approach Against Membership Inference Attacks

Matthew Chan[1], Aolin Ding[2]([✉]), Amin Hass[2], and Saman Zonouz[3]

[1] Rutgers University, New Brunswick, NJ, USA
[2] Security R&D, Accenture Labs, Accenture, Washington, DC, USA
a.ding@accenture.com
[3] Georgia Institute of Technology, Atlanta, GA, USA

Abstract. The proliferation of AI in Machine Learning as a Service (MLaaS) platforms has brought increased attention to the security and privacy of AI models, particularly within mobile edge computing environments. The ability to easily harness the analytical power of machine learning highlights privacy concerns when models are trained using sensitive data, as recent research has shown that models may retain and subsequently leak this information. Exposure of a participant's membership in the training set of a sensitive model, known as a Membership Inference Attack (MIA), can have severe personal and legal consequences. In this work, we explore using an empirical pruning approach to increase model robustness against privacy attacks like MIA, while simultaneously enhancing model efficiency-a crucial aspect for mobile edge and IoT deployments. Using our approach, we observe a moderate decrease in MIA recall with no comparable changes in MIA precision. With further refinements to our approach, we believe that model pruning can be a useful tool for protecting models against privacy attacks, as well as contributing to the security design of MLaaS applications in IoT and edge computing environments.

Keywords: Edge Computing · IoT Security and Privacy · DNN Pruning · Membership Inference Attack

1 Introduction

The rapid growth of Machine Learning - and in particular, Machine-Learning-as-a-Service (MLaaS), has allowed companies and researchers to access finely-tuned Deep Neural Networks (DNNs) and provide external parties with tools for training and deploying DNNs from scratch on their own data. We have seen this significant trend in mobile edge computing, where computational power is decentralized and brought closer to data sources [5,23]. In parallel, the explosion of network-connected mobile and embedded Internet-of-Things (IoT) devices has increased the capability to feed these models with high-resolution, context

---

M. Chan and A. Ding—Equally contributed as co-first authors.

W. Meng et al. (Eds.): ADIoT 2024, LNCS 15397, pp. 22–32, 2025.
https://doi.org/10.1007/978-3-031-85593-1_2

specific data [6]. Soon, DNNs will be deployed to these devices themselves, as a powerful analytical tool for converting raw, low-level data into high-level insights.

However, the popularization of MLaaS has brought attention to privacy concerns related to sensitive data used to train models, as any privacy violations can often expose sensitive personal, medical, or financial information. For example, it has been shown in [12,22,24] that anonymized data can leak information leading to deanonymization.

Research has shown that even with only black-box query access, DNNs pose the risk of leaking information about whether a data record was part of its training set - known as a Membership Inference Attack (MIA). This can be undesirable in many situations, for example facial recognition models trained using criminal databases [26] might allow an MIA to reveal an individual's criminal history. First highlighted by Shokri et al. [19], MIA analyzes information leaked by model responses to distinguish differences in model behavior between members and non-members of the training set.

In response, variety of defenses against MIA have also been proposed. Some techniques aim to combat MIA by improving model generalization [17,19], as MIA has some connections to overfitting the training data [27]. Other approaches modify model outputs to minimize MIA success [9,13].

In this paper, we explore empirical model pruning as a defense against MIA. Unlike other defense approaches, pruning offers to be a simpler and more flexible approach, offering DNNs additional advantages such as reduced computation and increased execution speed. This efficiency gain particularly benefits mobile edge and IoT devices with limited computational resources and power. And while we are not the first work [25,28] to utilize pruning to defend against privacy attacks, we use a novel dataset-focused approach rather than only considering the model or the aggregated training set.

This choice is based on insights gained from recent advancements in MIA, discussed further in Sects. 2 and 3. In Sect. 4, we present our methodology and preliminary experimental results. We discuss our findings and future avenues of research in Sect. 5, and conclude in Sect. 6.

2   Background and Related Work

2.1   Membership Inference Attacks (MIAs)

Membership Inference Attacks, first proposed by Shokri et al. [19], are attacks on the privacy of the data used to train a DNN. Using systematic querying of even a black-box model, it is possible for an attacker to determine whether or not a data point was a member of the training dataset. Formally, for a DNN $f(x) \rightarrow [0,1]^n$ classifying input x to an n-class confidence vector, membership inference attempts to build a binary classifier $A(x, f) \rightarrow \{0, 1\}$ for determining if a sample x is $\{1\}$ or isn't $\{0\}$ a member of f's training set.

Recent research has expanded on MIA techniques, showing that they can be conducted with less effort [17,27] using a single shadow model or simple thresholding, less information [4,11] with MIA possible when f outputs only a class label, and even greater precision [20,28] against robust DNN models. However, state-of-the-art MIA techniques which produce the most precise membership inferences [3,15,16] train an increasing number of (up to a thousand) shadow models. While computationally intense, producing a sufficient sample size of shadow models f'_{in} and f'_{out} for which sample x was or was not in the training set results in a high-quality membership inference.

2.2 Model Pruning

Model pruning is a popular method for model compression [8,21] due to its simplicity and ability to be performed on pre-trained models. This allows the use of large, high-accuracy neural networks on resource constrained devices such as those in IoT or Edge scenarios. This is achieved by removing extraneous model parameters in a structured or unstructured way, then fine-tuning the resulting model to maintain accuracy.

In addition to reducing model space and computational complexity, many recent pruning works have shown that different pruning metrics [14,18] can produce networks that are robust to adversarial examples. Comparatively less research has been done on the viability of pruning as a way to combat privacy threats, with [25] deriving and minimizing an adversarial gain function for pruning, and [28] using a posterior balancing loss during fine-tuning to mitigate confidence discrepancies that MIAs use.

3   Design and Methodology

In this section, we detail the design of our pruning framework, as illustrated in Fig. 1. Our goal is to increase the robustness of the model against MIA by pruning neurons corresponding to members of the training set that are vulnerable to MIA. Once identified, we use this information to prune neurons from the model to reduce the effectiveness of MIA. We describe how these samples are identified, and the pruning heuristic we investigate to reduce the effectiveness of MIA.

3.1   Identifying Vulnerable Samples

The vulnerability of a sample is dependent on a variety of factors including the rest of the training set, a model's parameters, and its training procedure. These influences are difficult to separate and quantify. Intuitively, the most straightforward way to establish a baseline for identifying vulnerable samples is by performing a membership inference attack on the model itself. However, the resource intensity of recent Bayesian-style approaches make them computationally infeasible. As a result, we sacrifice some utility in favor of efficiency in our experiments, using a threshold-based confidence vector attack which trains a single shadow model. Despite this, due to our iterative pruning approach which we describe below, we can show relative increases and decreases in MIA power, a trend that should be consistent across different MIA attacks.

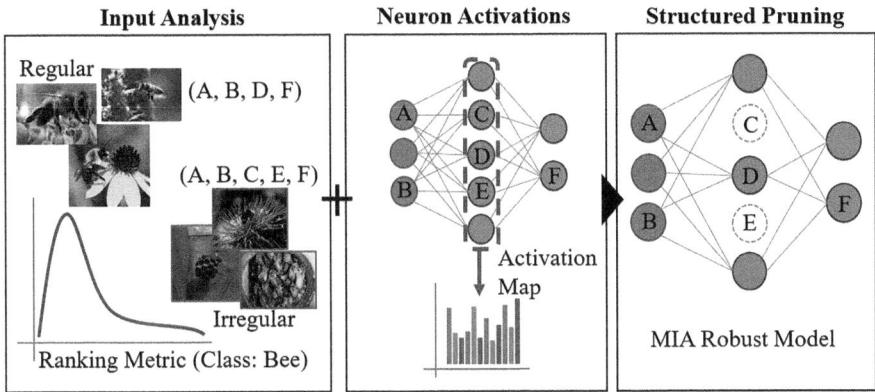

Fig. 1. A high-level overview of our pruning framework components, which include analyzing both the training dataset inputs and model behaviors (activation map) to perform MIA robust pruning.

The consideration of training set characteristics is important, as a sample's privacy risk is not only tied to the model, but also other data samples it was trained with. Recent MIA works have identified this connection between out-of-distribution data and privacy risk, but measure it with the aforementioned expensive approaches. For this reason, we investigate proxy metrics which may be suitable for identifying vulnerable samples in Sect. 5.

3.2 Pruning Approach

Because privacy attacks focus on a model's behavior on data, it makes greater sense to prune based on neuron activation rather than model weights. We conduct several iterative rounds of structured pruning, selecting the single layer with the greatest number of neurons that can be pruned at each round, as determined by our pruning heuristic. After each round, the model is fine-tuned with a retraining step to recover lost accuracy. The iterative structured approach prevents irrecoverable drops in model accuracy, and allows for computational speedups when compared to an unstructured pruning approach.

3.3 Pruning Heuristic

We investigate heuristic metrics for pruning based on neuron activations within the model when it performs inference on the vulnerable samples. We apply these heuristic on a per-layer basis, pruning the layer that has the greatest number of neurons that can be pruned.

The main pruning heuristic we have investigated so far, which we call MIA-Differential Pruning, leverages an existing statistic [8] used for magnitude activation pruning, known as Average Percentage of Zero Activations (APoZ). APoZ

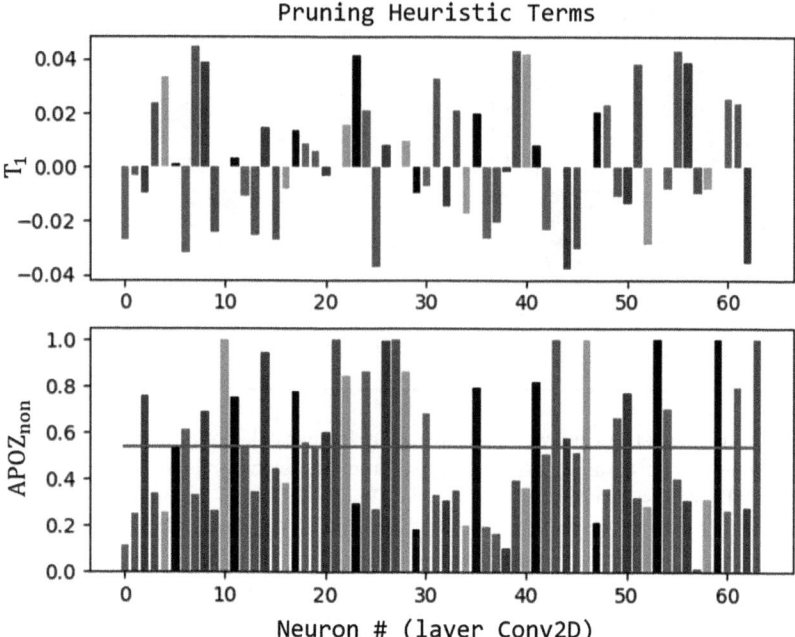

Fig. 2. Heuristic terms T_1 (top) and T_2 (bottom) on a layer with 64 neurons. Prunable neurons are positive in T_1 and above the line (average APoZ) for T_2.

is calculated based on the average number of zeros in the feature map of a neuron's output, calculated over a set of input samples. A high APoZ indicates that a neuron's output has many zeros (low activation), while a low APoZ indicates the neuron has few zeros (high activation). A useful corollary of APoZ is that neurons that output many zeros contribute little to the output of subsequent layers. We first partition the target model's training set into the set of identified vulnerable members and non-identified members. We then calculate the APoZ for both partitions for each neuron i in a layer with N neurons, denoted APoZ_{vuln}^i and APoZ_{non}^i. We compute the following two terms, where α is a scaling hyperparameter:

$$T_1^i = \mathrm{APoZ}_{non}^i \ - \ \mathrm{APoZ}_{vuln}^i$$

$$T_2^i = \alpha \cdot \mathrm{APoZ}_{non}^i \ - \ \frac{1}{N} \sum_{i=1}^{N} \mathrm{APoZ}_{non}^i$$

A visualization of application of these terms can be seen in Fig. 2. When both terms are positive for neuron i, this neuron is prunable. When T_1^i is positive, it implies that vulnerable samples have lower APoZ on average (equivalent to more non-zero activations). Thi When T_2^i is positive, this means that neuron i has above average APoZ relative to the average APoZ across the entire layer; this indicates that this neuron is not overly important. We can use α to scale the importance threshold up ($\alpha < 1$) or down ($\alpha > 1$). In our experiments we found $\alpha \geq 1.1$ to be necessary, otherwise very few parameters were pruned.

The benefit of this heuristic is that it is easy to compute and use, and balances robustness (T_1) and accuracy (T_2). However, the process of statistical averaging may result in the loss of more nuanced vulnerability information. We further discuss heuristic design choices and evaluation in Sect. 5.

4 Evaluation

4.1 Experimental Setup

We conduct our experiments on the CIFAR-10 dataset, using a 4-layer model with around 1 million parameters as well as a well-generalized VGG-16 model containing around 15 million parameters. The models were trained until the point where testing accuracy did not increase for 20 consecutive epochs, on half of the 60k CIFAR-10 dataset to accommodate shadow model training. The original and post-pruning test/train accuracies are shown in Table 1. We conduct 10 rounds of iterative pruning to observe MIA trends, followed by 20 additional epochs of retraining after each pruning iteration.

Table 1. Accuracy comparisons for the original trained model, and the resulting final model after all pruning iterations.

(Train/Test Acc.)	4-Layer	VGG16
Original	99.5%/68.3%	86.6%/77.1%
Magnitude	100%/65.2%	99.8%/71.8%
Differential	100%/54.3%	99.4%/71.8%

In our experiments, we measure the effectiveness of our MIA attack using Precision and Recall evaluated over a balanced subset of training and non-training samples.

4.2 Baseline Comparison

To set a baseline for our approach, we iteratively prune a model using activation magnitude pruning. This process simply prunes neurons with high APoZ. Figure 3 shows the results of MIA over successive pruning iterations. Noticably, in both models magnitude pruning increases the recall of MIA, meaning that a greater number of training samples are correctly inferred as members. However, MIA precision does not decline correspondingly, indicating the attack is overall stronger.

For the VGG16 model, the initial recall value is much lower; we attribute this to its dropout regularization during training. Interestingly, these benefits

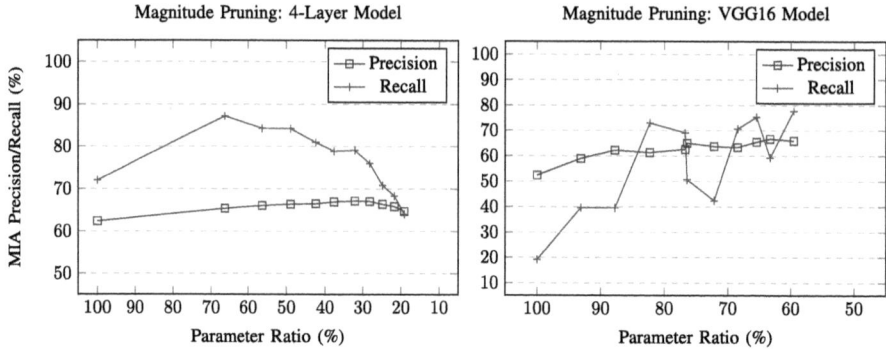

Fig. 3. MIA results against Magnitude Pruning for the 4-layer and VGG16 models.

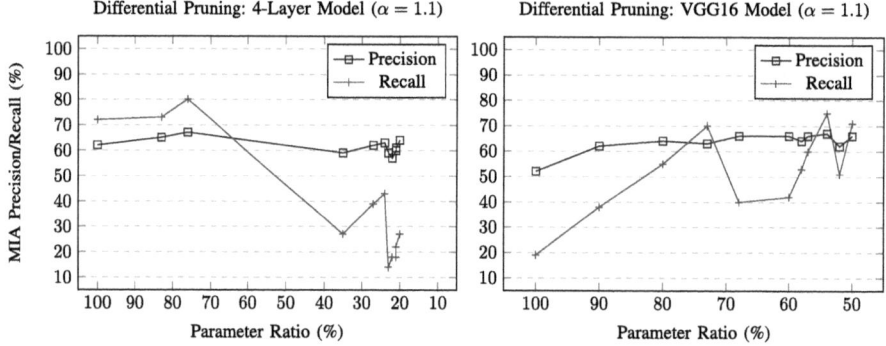

Fig. 4. MIA results against MIA-Differential Pruning for the 4-layer and VGG16 models.

seem to be lost when the model is pruned, despite retraining. We hypothesize that the sharp changes in recall on VGG16 are due to the increased number of layers relative to the single layer that we prune.

4.3   MIA-Differential Pruning Heuristic

Next, we test our pruning approach against MIA. Figure 4 shows an equivalent experiment to the baseline, with only the pruning heuristic changed. We vary the value of α between 1 and 1.3 in increments of 0.1, finding that the value should be greater than 1 in order for the heuristic to select a sufficient number of neurons to prune. However we see less change as α increases with respect to its MIA performance, limited to a slight increase in parameters pruned and lower accuracy as a result. Unlike the baseline attack, in the 4-Layer model we can see strong reductions in MIA recall, and a slightly slower increase in MIA recall for the VGG16 model. In both cases, precision remains stable throughout pruning. As a result, the MIA attack effectiveness is overall weaker, though only slightly in the case of VGG16.

5   Discussion: Limitations and Future Work

5.1   Limitations

MIA Challenges. In its current form, this work has several limitations related to MIA itself. As described in Sect. 3, the most precise MIA attacks are computationally impractical for use in this defense mechanism, but still pose a threat from a determined adversary (Carlini et al. [3] describe an offline attack that amortizes the computational investment). Even so, threshold-based MIA attacks have a non-negligible computational cost and suffer from lower precision as detailed in related work. While still worth investigating, this issue gives rise to the next set of challenges.

Evaluating Heuristics. Due to the pipelined nature of our described framework, steps earlier in the process have cascading effects. We seek to use MIA as a "ground truth" for sample vulnerability because it in theory true by definition. However in practice, inconsistent or inaccurate inference results make it challenging to properly evaluate pruning heuristics. Despite these challenges, we are still able to see a trend due to the iterative pruning rounds, which offer a relative comparison which is slightly more stable.

5.2   Future Work

Proxy Metrics. Efficiently computable proxies resolve the computational costs associated with discerning vulnerable samples. Jiang et al. [10] present a potential proxy in their Consistency score, which characterizes the structural regularity of the dataset. The analytical computation of the C-score closely mirrors the Bayesian sampling of newer MIA attacks, and show high-fidelity. Additionally, [2] look at the variance of gradients (VoG) in order to rank samples in terms of learning difficulty. In the continuance of this work, we plan to explore these promising proxy metrics as part of a complete evaluation.

Privacy Attack Extension. Another promising direction for future work in this area lies in applying pruning to other privacy attacks. The original introduction of model pruning was as a model compression technique to reduce the size and computational complexity of rapidly-growing DNNs. However, with a few exceptions [25,28], we believe that model pruning has not been sufficiently explored in defending against privacy attacks. With recent research [7] implying that some degree of training data memorization may be necessary to achieve optimal model generalization, privacy may become unavoidable tradeoff with model accuracy [1]. Pruning offers a compromise, where vulnerable samples which may be important for accuracy don't have to be removed from the dataset. Instead, intelligent pruning may force a model to find a different internal representation which leaks less information.

6   Conclusion

In this paper, we present preliminary results for an empirical pruning approach to defend against membership inference attacks. This exploration innovatively utilizes a sample-focused approach, specifically targeting training records vulnerable to MIA. Our initial results show moderate relative improvement against MIA, reducing attack recall without increasing attack precision. This work demonstrates some of the potential model pruning has to combat privacy attacks, which seek to extract latent information from a model. Our findings suggest that with model pruning, it is promising to enhance both security and efficiency in ML applications deployed at the edge.

Acknowledgements. The work presented in this paper was mainly done while the author M. Chan was a research intern at Accenture Labs. This work was supported by the National Science Foundation (NSF) Cyber-Physical Systems (CPS), Secure and Trustworthy Cyberspace (SaTC) Programs, Accenture, and Department of Energy (DOE) Award DE-OE0000780 Cyber Resilient Energy Delivery Consortium, and DE-CR0000056 Physics-Aware and Automated Vulnerability Discovery and Patching in Heterogeneous Distributed Energy Resources against Cyber Attacks.

References

1. Abadi, M., et al.: Deep learning with differential privacy. In: Proceedings of the 2016 ACM SIGSAC Conference on Computer and Communications Security, pp. 308–318 (2016)
2. Agarwal, C., D'souza, D., Hooker, S.: Estimating example difficulty using variance of gradients. In: Proceedings of the IEEE/CVF Conference on Computer Vision and Pattern Recognition, pp. 10368–10378 (2022)
3. Carlini, N., Chien, S., Nasr, M., Song, S., Terzis, A., Tramer, F.: Membership inference attacks from first principles. arXiv preprint arXiv:2112.03570 (2021)
4. Choquette-Choo, C.A., Tramer, F., Carlini, N., Papernot, N.: Label-only membership inference attacks. In: International Conference on Machine Learning, pp. 1964–1974. PMLR (2021)
5. Ding, A., Hass, A., Chan, M., Sehatbakhsh, N., Zonouz, S.: Resource-aware DNN partitioning for privacy-sensitive edge-cloud systems. In: International Conference on Neural Information Processing, pp. 188–201. Springer, Cham (2023)
6. Ding, A., Murthy, P., Garcia, L., Sun, P., Chan, M., Zonouz, S.: Mini-me, you complete me! Data-driven drone security via DNN-based approximate computing. In: Proceedings of the 24th International Symposium on Research in Attacks, Intrusions and Defenses, pp. 428–441 (2021)
7. Feldman, V., Zhang, C.: What neural networks memorize and why: discovering the long tail via influence estimation. In: Advances in Neural Information Processing Systems, vol. 33, pp. 2881–2891 (2020)
8. Hu, H., Peng, R., Tai, Y.W., Tang, C.K.: Network trimming: a data-driven neuron pruning approach towards efficient deep architectures. arXiv preprint arXiv:1607.03250 (2016)

9. Jia, J., Salem, A., Backes, M., Zhang, Y., Gong, N.Z.: Memguard: defending against black-box membership inference attacks via adversarial examples. In: Proceedings of the 2019 ACM SIGSAC Conference on Computer and Communications Security, pp. 259–274 (2019)
10. Jiang, Z., Zhang, C., Talwar, K., Mozer, M.C.: Characterizing structural regularities of labeled data in overparameterized models. arXiv preprint arXiv:2002.03206 (2020)
11. Li, Z., Zhang, Y.: Membership leakage in label-only exposures. In: Proceedings of the 2021 ACM SIGSAC Conference on Computer and Communications Security, pp. 880–895 (2021)
12. Narayanan, A., Shmatikov, V.: Robust de-anonymization of large sparse datasets. In: 2008 IEEE Symposium on Security and Privacy (SP 2008), pp. 111–125. IEEE (2008)
13. Nasr, M., Shokri, R., Houmansadr, A.: Machine learning with membership privacy using adversarial regularization. In: Proceedings of the 2018 ACM SIGSAC Conference on Computer and Communications Security, pp. 634–646 (2018)
14. Phan, H., Yin, M., Sui, Y., Yuan, B., Zonouz, S.: CSTAR: towards compact and structured deep neural networks with adversarial robustness. In: Proceedings of the AAAI Conference on Artificial Intelligence, vol. 37, pp. 2065–2073 (2023)
15. Rezaei, S., Liu, X.: On the difficulty of membership inference attacks. In: Proceedings of the IEEE/CVF Conference on Computer Vision and Pattern Recognition, pp. 7892–7900 (2021)
16. Sablayrolles, A., Douze, M., Schmid, C., Ollivier, Y., Jégou, H.: White-box vs black-box: Bayes optimal strategies for membership inference. In: International Conference on Machine Learning, pp. 5558–5567. PMLR (2019)
17. Salem, A., Zhang, Y., Humbert, M., Berrang, P., Fritz, M., Backes, M.: ML-leaks: model and data independent membership inference attacks and defenses on machine learning models. arXiv preprint arXiv:1806.01246 (2018)
18. Sehwag, V., Wang, S., Mittal, P., Jana, S.: Hydra: pruning adversarially robust neural networks. In: Advances in Neural Information Processing Systems, vol. 33, pp. 19655–19666 (2020)
19. Shokri, R., Stronati, M., Song, C., Shmatikov, V.: Membership inference attacks against machine learning models. In: 2017 IEEE Symposium on Security and Privacy (SP), pp. 3–18. IEEE (2017)
20. Song, L., Mittal, P.: Systematic evaluation of privacy risks of machine learning models. In: 30th USENIX Security Symposium (USENIX Security 21), pp. 2615–2632 (2021)
21. Sui, Y., Yin, M., Xie, Y., Phan, H., Aliari Zonouz, S., Yuan, B.: Chip: channel independence-based pruning for compact neural networks. In: Advances in Neural Information Processing Systems, vol. 34, pp. 24604–24616 (2021)
22. Tang, M., et al.: Modelguard: information-theoretic defense against model extraction attacks. In: 33rd USENIX Security Symposium (Security 2024) (2024)
23. Tang, M., et al.: Fade: enabling federated adversarial training on heterogeneous resource-constrained edge devices. arXiv preprint arXiv:2209.03839 (2022)
24. Waheed, N., He, X., Ikram, M., Usman, M., Hashmi, S.S., Usman, M.: Security and privacy in IoT using machine learning and blockchain: threats and countermeasures. ACM Comput. Surv. (CSUR) **53**(6), 1–37 (2020)
25. Wang, Y., et al.: Against membership inference attack: pruning is all you need. arXiv preprint arXiv:2008.13578 (2020)
26. Wu, X., Zhang, X.: Automated inference on criminality using face images. arXiv preprint arXiv:1611.04135 pp. 4038–4052 (2016)

27. Yeom, S., Giacomelli, I., Fredrikson, M., Jha, S.: Privacy risk in machine learning: analyzing the connection to overfitting. In: 2018 IEEE 31st Computer Security Foundations Symposium (CSF), pp. 268–282. IEEE (2018)
28. Yuan, X., Zhang, L.: Membership inference attacks and defenses in neural network pruning. In: 31st USENIX Security Symposium (USENIX Security 22), pp. 4561–4578 (2022)

A Conflict-Aware Active Automata Learning Approach for BLE Device Status Machine Construction

Jian Xu, Long Yin(✉)(iD), Heqiu Chai, Zhongsheng Wang, and Chunyu Liu

Software College, Northeastern University, Shenyang 110169, China
2110499@stu.neu.edu.cn

Abstract. To identify potential security vulnerabilities in BLE devices, existing automated detection methods, including state machine learning, differential testing, and fuzz testing, face limitations. These approaches struggle with constrained interaction data and lack of timely feedback, making it difficult to accurately learn complex state machine models. Current fuzz testing methods are often restricted to shallow protocol states, overlooking deeper critical behaviors, and the impact of wireless communication can introduce uncertainty in device responses. To address these limitations, this paper introduces a conflict-aware active learning method. Based on the Minimally Adequate Teachers (MAT) framework, we designed a cache tree, discrimination tree, observation tree, and probabilistic model. Using the cache tree, we constructed an initial state machine for the BLE protocol. The discrimination tree generated an active learning query sequence, while the observation tree and probabilistic model were combined to create conflict detection and resolution algorithms. Finally, experiments demonstrated the effectiveness and robustness of the proposed method.

Keywords: Bluetooth Low Energy · grey-box fuzzing · active automata learning · state coverage

1 Introduction

Bluetooth Low Energy (BLE) is a crucial enabler for the development of Internet of Things (IoT) technologies, offering low power consumption, cost-effectiveness, and convenience. It has been widely adopted across billions of smart devices in various fields, including personal health, smart homes, and industrial control. However, as BLE devices are deployed on a large scale, security issues have become increasingly prominent, drawing attention from both academia and industry. Research has revealed serious security vulnerabilities in the protocol stacks of mainstream BLE chip manufacturers. For instance, Armis discovered a buffer overflow vulnerability in Texas Instruments (TI) BLE chip protocol stacks that allows attackers to execute arbitrary code or shut down devices by sending

© The Author(s), under exclusive license to Springer Nature Switzerland AG 2025
W. Meng et al. (Eds.): ADIoT 2024, LNCS 15397, pp. 33–52, 2025.
https://doi.org/10.1007/978-3-031-85593-1_3

malformed packets, affecting about 70% of enterprise access point devices. Additionally, Garbelini and colleagues exposed vulnerabilities in the protocol stacks of multiple BLE chip manufacturers, which could bypass existing security mechanisms and potentially cause device crashes or deadlocks, impacting over 480 different BLE devices.

These findings highlight systemic issues in BLE protocol implementations, where chip manufacturers have insufficiently considered security, resulting in discrepancies between protocol stacks and design standards. Although the Bluetooth Special Interest Group introduced security mechanisms in BLE 4.0 and made significant improvements in BLE 4.2 and BLE 5.0, real-world security issues remain largely unresolved. These problems often stem from misunderstandings or neglect of standard documentation by chip manufacturers, leading to inconsistencies in protocol stack implementations. Additionally, physical limitations of devices, such as constraints on computational power and input capabilities, may prevent full support for all security features.

To effectively identify potential security vulnerabilities in BLE devices, there is an urgent need for efficient security testing techniques. Fuzz testing, an automated software testing approach, inputs a large number of randomly generated anomalous test cases into a target system and monitors the system's responses to these inputs to uncover potential vulnerabilities. The effectiveness of fuzz testing in discovering protocol vulnerabilities has been widely validated [2–4]. As BLE technology continues to expand in the IoT domain, developing efficient fuzz testing methods specifically for BLE devices has become increasingly urgent and important.

Our main contributions are as follows:

1. We proposes a conflict-aware active learning method based on the MAT (Minimally Adequate Teachers) framework to efficiently construct BLE device state machines, which integrates the cache trees, discrimination trees, observation trees, and a probability model.
2. The proposed method uses the discrimination trees to select the most informative input sequences and optimize the query strategies during the learning process. It also reuses query-response pairs from the cache tree to significantly reduce query counts and data volume, thereby improving learning efficiency.
3. The proposed method employs the observation trees to manage state transitions and their frequencies observed during real interactions, effectively tracking and quantifying uncertainty in the learning process. And the probability model calculates the likelihood of different states and transitions, resolving conflicts and dynamically updating the BLE device state machine.

2   Related Works

As an emerging IoT communication technology, the security of BLE has garnered significant attention. In recent years, researchers have identified multiple critical vulnerabilities in BLE protocol stacks and implementations through various methods, including manual analysis, static modeling, and dynamic testing.

Early efforts primarily relied on the expertise of security researchers, employing traditional methods such as code auditing, hardware analysis, and firmware reverse engineering to manually identify potential vulnerabilities. Seri and Vishnepolsky [1] conducted a comprehensive audit of the source code, ROM code, and library code of TI's BLE protocol stack, uncovering a memory corruption vulnerability during the parsing of broadcast packets that could lead to remote code execution. By extracting and analyzing the BLE chip firmware from various manufacturers' access point products, they effectively exploited these vulnerabilities in practice, ultimately gaining control over enterprise-level WiFi networks.

Static modeling analysis methods involve constructing formal models based on BLE protocol specifications and then using model checking or reasoning techniques to identify potential security vulnerabilities. Wu et al. [5] developed a formal model of the BLE protocol using the formal modeling language ProVerif, systematically analyzing its security. They performed formal modeling and verification of the BLE reconnection process, uncovering two critical design flaws in the BLE protocol specification and proposing the BLESA attack based on these flaws. The BLESA attack enables an attacker to impersonate a previously paired server device and inject forged data into the client, thereby bypassing authentication mechanisms. In subsequent work, Wu et al. [6] expanded the coverage of their formal model to create a comprehensive BLE protocol model that includes key negotiation and data transmission phases, as well as the classic Bluetooth, BLE, and BLE Mesh protocols. Using this integrated model, they not only replicated previously identified vulnerabilities but also discovered several new security issues.

Fuzz testing is a widely used technique for software security, achieving significant success in discovering numerous critical vulnerabilities [7]. However, applying fuzz testing to Bluetooth protocols presents several challenges. First, triggering Bluetooth protocol vulnerabilities often requires injecting a series of packets while adhering to strict timing constraints, whereas traditional fuzz testing tools are more suited for vulnerabilities triggered by single inputs. Additionally, the implementation details of most commercial Bluetooth protocol stacks are proprietary, making it difficult to use gray-box or white-box fuzz testing methods based on code coverage or symbolic execution [8,9]. Furthermore, wireless communication is characterized by randomness, packet loss, and retransmission, complicating the differentiation between normal and anomalous behavior in fuzz testing [10]. Finally, the isolation between the link layer and host layer in Bluetooth protocol stacks poses further challenges for comprehensive security testing [11].

In response to these challenges, security researchers have proposed various solutions and tools. Defensics [12] explored fuzz testing for classic Bluetooth but did not incorporate optimization strategies to enhance testing efficiency, and it can only inject a limited number of packet fields after peripheral devices are connected. Ruge et al. [13] developed the Frankenstein testing framework, which enables deep and systematic testing of Bluetooth chip firmware without requiring physical devices, discovering three zero-day vulnerabilities in

Broadcom and Cypress Bluetooth stacks. Garbelini et al. [11] designed a BLE controller firmware that abstracts the timing and retransmission requirements between central and peripheral devices, allowing the host to manipulate all fields of link layer packets, thus providing foundational support for BLE fuzz testing. They constructed a universal BLE protocol state machine model based on the BLE specifications and used model-based fuzz testing to reveal security issues in low-energy Bluetooth devices, identifying 11 new vulnerabilities across multiple mainstream BLE chips that affect over 480 different devices.

Park et al. [14] designed a fuzz testing tool targeting the L2CAP layer, using state-guided and core field mutation techniques, successfully detecting five zero-day vulnerabilities in actual Bluetooth devices. Pferscher et al. [15] addressed the challenges of black-box environments by employing state machine learning to create state machine models of target BLE devices, enhancing fuzz testing coverage and effectiveness through automatically inferred behavioral models. Karim et al. [16] developed the BLEDiff automated differential testing framework to tackle issues in BLE protocol implementations. By leveraging finite state machines and active state machine learning methods, BLEDiff automatically extracts protocol state machines from BLE implementations and effectively identifies anomalous behaviors through the partitioning and synthesis of multiple sub-protocols. Testing on 25 commercial devices revealed 13 different anomalous behaviors, including 10 exploitable security vulnerabilities. Shu et al. proposed a automated blackbox fuzz testing framework of IoT network protocols called IoTInfer, which is guided by finite state machine inference. It is better at eliciting different types of responses from the fuzzing targets than the other two state-of-the-art blackbox IoT device fuzzing tools IoTFuzzer [20] and Snipuzz [21].

However, these tools and methods also exhibit several limitations:

1. Inefficiency and Lack of Portability: The effectiveness of many tools relies heavily on the deep domain knowledge and labor-intensive manual operations of security researchers, which is not only time-consuming but also inefficient. Additionally, these tools are often designed for specific chipsets, lacking portability and making them unsuitable for direct application on devices from different manufacturers.

2. Insufficient Consideration of Implementation Details: Analysis using predefined protocol models tends to abstract and idealize specifications, making it difficult to capture the detailed flaws that may exist in actual implementations. This limitation hinders their ability to discover vulnerabilities effectively.

3. Inadequate Consideration of Wireless Communication Characteristics: There is a lack of deep consideration for the differences between wireless communication security testing and traditional security testing. The unreliability of wireless communication and the resulting non-deterministic outcomes can significantly reduce the efficiency of methods based on active learning.

3   BLE Device State Machine Generation Method

In Sect. 3.2, this paper presents the standard state machine model for the parameter exchange process before connection establishment in the BLE protocol. However, this model only describes the expected behavior that BLE devices should follow under ideal conditions. In real-world scenarios, due to variations in implementations by different manufacturers and factors such as wireless interference, the actual behavior of specific BLE devices often deviates from the standard specification. Therefore, it is crucial to accurately analyze and test specific BLE devices by constructing state machine models that precisely describe their actual state transitions and outputs during operation.

In this paper, the state machine learned from real device interactions is referred to as the BLE device state machine, which differs from the standard BLE state machine described in Sect. 3.2. This section will introduce the conflict-aware active learning method proposed in this study, which is designed to efficiently and accurately construct the state machine model from actual BLE device interactions.

3.1   Overall Framework

The overall framework of the conflict-aware active learning method is shown in Fig. 1. This method consists of three main processes: initialization, active learning, and conflict resolution. The core of the method lies in four components: the cache tree, discrimination tree, observation tree, and probabilistic model. These components work closely together to complete the construction of the BLE device state machine.

(1) **Initialization Process.** In the initialization phase, the method constructs an initial state machine based on prior knowledge of the BLE protocol, which is stored in the cache tree. This state machine describes the typical operating states and state transitions of the BLE device under default configurations, providing guidance for the subsequent active learning process. At the same time, an initial discrimination tree is also constructed to generate query sequences.

(2) **Active Learning Process.** In the active learning phase, the method uses the discrimination tree to generate query sequences. The discrimination tree is designed to generate query sequences that can most effectively differentiate between the currently known states. For the generated query sequences, the cache tree is first searched to see if there is a reusable query-response path. If a reusable path exists, the query is sent with a certain probability to verify the accuracy of the reusable path. If no reusable path is found in the cache tree, a new query sequence is executed. The query-response pairs are recorded in the observation tree. After updating the observation tree, the method checks whether the newly added response matches the expected response of the corresponding state in the cache tree. If they match, no further action is required. If the newly added response in the observation tree does not match the expected response of the corresponding state in the cache tree, a conflict is generated.

(3) Conflict Resolution Process. Once a conflict is detected between a new response in the observation tree and the cache tree, the method utilizes statistical data from the observation tree and employs the probabilistic model to calculate and compare the probabilities of each possible response. If a candidate solution has a probability that exceeds a predefined confidence threshold, it is selected as the resolution for the conflict, and the cache tree is updated accordingly.

The active learning process is repeated until a predetermined stopping condition is met, such as the number of query rounds or a time limit. Ultimately, the algorithm outputs the learned state machine for the target BLE device.

3.2   BLE Protocol State Machine Initialization Based on Cache Tree

The Cache Tree is a key data structure in the conflict-aware active learning method proposed in this paper. It is used to store the initial BLE state machine as well as the query sequences and corresponding state transition paths generated during the execution of the method. The core function of the cache tree is to encode the finite prefixes of the BLE state machine, allowing for efficient reuse of previous queries and state transitions.

Definition 1. *(**Cache Tree**) The cache tree is a rooted tree $T = (V, E)$, where V is the set of nodes and $E \subseteq V \times V$ is the set of edges. Each node $v \in V$ stores the current set of input packets $in(v) \subseteq I$, the set of output packets $out(v) \subseteq O$, and the corresponding protocol state $state(v)$. Here, I and O represent the entire set of input and output packets, respectively. Each edge $e = (u, v) \in E$ represents a transition from node u (with state $state(u)$) to node v, triggered by sending the input packet set $in(v)$ and receiving the output packet set $out(v)$ from the device.*

There is an empty root node $r \in V$, where $in(r) = out(r) = \emptyset$ and $state(r)$ is an empty state. All initial state nodes are direct children of the empty root node r. For any node $v \in V$, the input packet sequence from the root node to v can be retrieved by traversing the parent edges, forming the sequence $path(v)$, where $\langle \rangle$ denotes the empty sequence, and $\prod$ represents the concatenation operator that combines multiple sequences into a longer one. The sequence $\pi(v)$ is the collection of all edges from the root node r to node v.

$$path(v) = \langle \rangle \cdot \prod_{(u,v) \in \pi(v)} in(v) \qquad (1)$$

In formal terms, the cache tree encodes a finite prefix of a Mealy-type BLE protocol state machine $M = (S, s_0, I, O, \delta, \lambda)$. Specifically, for each child node v, $path(v)$ corresponds to a finite prefix path of the state machine M, and $state(v)$ represents the state reached by this path.

In the conflict-aware active learning algorithm, the initialization of the BLE state machine serves as the starting point for the entire learning process. Using the prior knowledge from BLE protocol specifications, an initial BLE state machine $M = (S, s_0, I, O, \delta, \lambda)$ is constructed and encoded into the cache tree $T = (V, E)$, following the process described below.

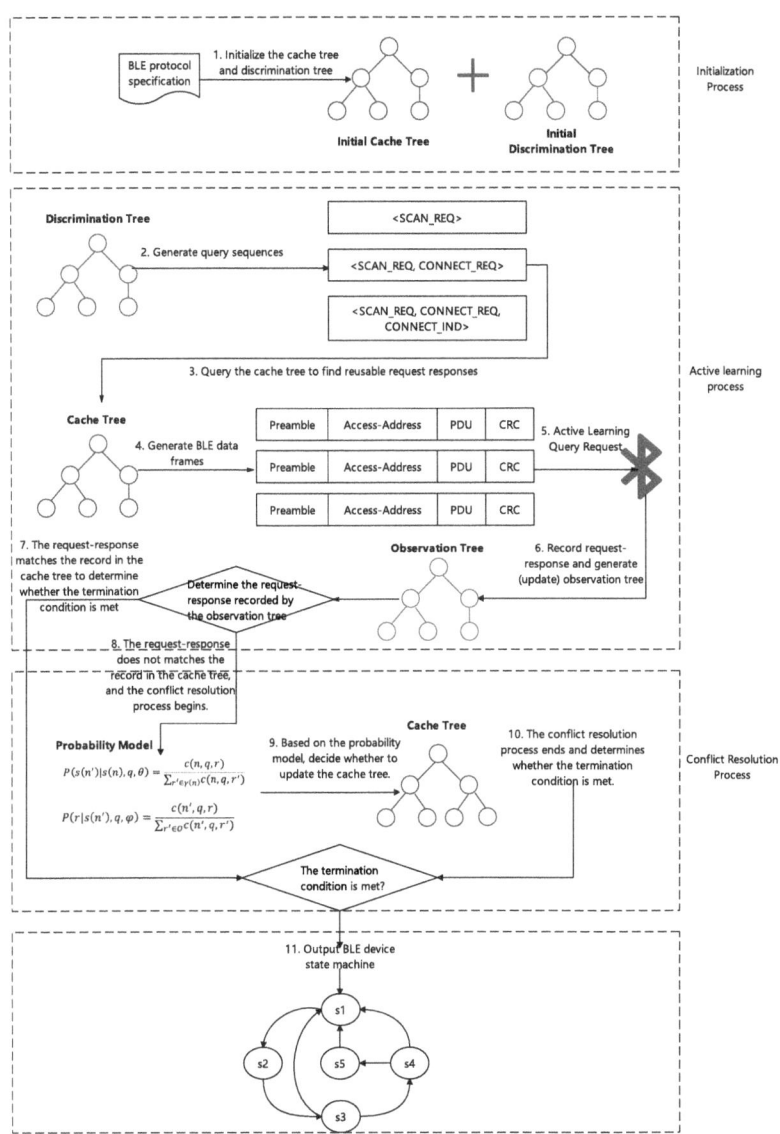

Fig. 1. The conflict-aware active automata learning process.

(1) Create the root node $r \in V$ such that $in(r) = out(r) = \emptyset$ and $state(r) = \emptyset$.

(2) For each initial state $s \in s_0$, create a child node $v_s \in V$ as a direct child of r, such that: $in(v_s) = out(v_s) = \emptyset$, $state(v_s) = s$.

(3) For each state $s \in S$ and input $i \in I$, if $\delta(s, i) = s'$ and $\lambda(s, i) = o$, then create an edge $(v_s, v_{s'}) \in E$, such that: $in(v_{s'}) = \{i\}$, $out(v_{s'}) = \{o\}$, $state(v_{s'}) = s'$.

Through the above iteration, the algorithm traverses all the states and state transitions in the BLE state machine M, encoding them into the cache tree T. During the subsequent active learning process, the cache tree will be continuously expanded to learn the complete state machine of the target BLE device.

3.3   Active Learning Query Based on Discrimination Tree

The Discrimination Tree is constructed based on the state transitions and query response information defined in the BLE protocol specification. Its purpose is to generate query sequences that can maximally distinguish the currently known states, thereby enhancing the query efficiency of active learning.

Definition 2. (Discrimination Tree) *The Discrimination Tree is a tree structure $T = (N, E)$, where N is the set of nodes and E is the set of directed edges. Each node $n \in N$ contains two components: U_n, which is the set of states corresponding to that node, and T_n, which is a list of triples (q, r, n'), where q is the data received by the target device, r is the response of the target device to q, and n' is the child node generated based on r. The set E is the collection of directed edges, with each edge $e \in E$ corresponding to n' in the triple (q, r, n') from a node n. There exists a root node n_r whose state set U_{n_r} contains all possible states. All leaf nodes n are single-state nodes, that is, $|U_n| = 1$.*

The process of generating query sequences using the Discrimination Tree is as follows: starting from the root node n_r, traverse the tree. For the current node n, arbitrarily select a triple (q, r, n') and add q to the current query sequence. Then, take n' as the next node to traverse, repeating the above process until a leaf node is reached, thus forming a query sequence π from the root node to the leaf node. Backtrack to the upper-level node to find new paths and repeat the above process, generating all possible query sequences $\Pi = \pi_1, \pi_2, \ldots, \pi_m$, as shown in Algorithm 1.

After obtaining a set of query sequences Π, the active learning process begins. First, the query sequences interact with the cache tree $T_c = (V, E_c)$, and actual queries are performed to obtain responses from the target device. For each query sequence $\pi_i \in \Pi$, the cache tree T_c is searched to check if there is a reusable request-response path. Specifically, starting from the root node of the cache tree, for each query instruction q in π_i, check if there exists a node v such that $in(v) = q$. If such a v exists, a probability p is used to decide whether to actually execute query q. If query q is executed with probability p, the response r' from the target device is obtained, and the observation tree is updated. If $r' = out(v)$, it indicates that the path from node v can be reused, and the traversal continues from node v; if $r' \neq out(v)$, a conflict is detected, and the conflict is resolved using a probabilistic model. Based on the conflict resolution result, it is decided whether to create a new node in T_c. If query q is not executed, the traversal directly continues from node v. If no such v exists, query q is executed, the response r' is obtained, the observation tree is updated, and the conflict is resolved. After executing each π_i and updating both the observation

---

Algorithm 1. Generate query sequence using discrimination tree

---

Input: Discrimination Tree $T = (N, E)$, Root node n_r
Output: Query sequence set Π
1: $\Pi \leftarrow \emptyset$ ▷ Initialize the query sequence set
2: **procedure** GENERATEQUERY$(n, sequence)$
3:     **if** $|U_n| = 1$ **then**   ▷ Check if it is a leaf node
4:         $\Pi \leftarrow \Pi \cup \{sequence\}$ ▷ Add a sequence to the query sequence
5:     **else**
6:         **for all** $(q, r, n') \in T_n$ **do**   ▷ Traverse all triples of the current node
7:             $newSequence \leftarrow sequence + [q]$ ▷ Add new query to the sequence
8:             GENERATEQUERY$(n', newSequence)$ ▷ Recursive call to process child
    nodes
9:         **end for**
10:     **end if**
11: **end procedure**

---

tree and cache tree, the learning of the current state and state transitions of the target device is complete. By traversing the entire Π and updating the observation and cache trees, the learning process of the target device's state machine is continuously optimized. This process is detailed in Algorithm 2.

3.4   Conflict Resolution Based on Observation Tree and Probabilistic Model

In the process of learning the BLE device state machine, if the observed device behavior is inconsistent with the current state machine model, a conflict arises. The presence of a conflict indicates that there is a discrepancy between the current state machine model and the actual device behavior, necessitating modifications and updates to the model.

Definition 3. *(Conflict) Let the current learned BLE device state machine model be $M = (S, s_0, I, O, \delta, \lambda)$. During the learning process, if there exists a state $s \in S$, a query $q \in I$, and a response $r \in O$, such that:*

1. *$\delta(s, q)$ is defined, meaning that sending the query q from state s is valid.*
2. *The observed actual response $r \neq \lambda(s, q)$.*

Then, we say that when the query q is sent from state s, the observed response r is in conflict with the current model M.

To resolve conflicts, the method introduces an observation tree and a probabilistic model.

Definition 4. *(Observation Tree) The observation tree is a tree-like data structure $O = (N, E, \Sigma, L, \gamma)$, where:*

- *N is the set of nodes in the observation tree,*

Algorithm 2. Active learning query algorithm

Input: cache tree $T_c = (V, E_c)$, query sequence set Π, query execution probability p
Output: updated cache tree T_c, updated observation tree T_o
1: **procedure** PROCESSQUERYSEQUENCES(T_c, Π, p)
2:     **for** $\pi_i \in \Pi$ **do**   ▷ Traverse all query sequences
3:         **for** $q \in \pi_i$ **do**   ▷ Process each query instruction in the sequence
4:             $v \leftarrow$ FINDNODE(T_c, q)
5:             **if** $v \neq$ null **then** ▷ If the corresponding query is found in the cache tree
6:                 $execute \leftarrow p$
7:                 **if** $execute$ **then**   ▷ Determine whether to execute the query with
probability p
8:                     $r' \leftarrow$ EXECUTEQUERY(q)
9:                     UPDATEOBSTREE(T_o, q, r')
10:                     **if** $r' = $ out(v) **then**   ▷ If the device response is consistent with
the cache
11:                         Continue
12:                     **else**   ▷ If the response is inconsistent, resolve the conflict
13:                         RESOLVECONFLICT(T_c, T_o, v, r')
14:                     **end if**
15:                 **else**   ▷ Do not execute the query, continue to use the cache data
16:                     Continue
17:                  **end if**
18:             **else**   ▷ Query not found in cache
19:                 $r' \leftarrow$ EXECUTEQUERY(q)
20:                 UPDATEOBSTREE(T_o, q, r')
21:                 RESOLVECONFLICT(T_c, T_o, v, r')
22:              **end if**
23:         **end for**
24:     **end for**
25: **end procedure**

- $E \subseteq N \times \Sigma \times N$ is the set of edges,
- Σ is the finite set of input queries,
- L is the finite set of output responses, and
- $\gamma : N \to 2^L$ associates each node $n \in N$ with a response set $\gamma(n) \subseteq L$, representing the set of responses that may be observed in the state corresponding to node n.

Here, 2^L denotes the power set of L, which is the set of all subsets of L.

For any node $n \in N$ in the observation tree, let $s(n)$ represent the state corresponding to node n. Each edge $(n, q, n') \in E$ represents a transition from the state $s(n)$ corresponding to node n, triggered by sending the query $q \in \Sigma$, which may lead to a transition to the state $s(n')$ corresponding to node n'.

Each node n also maintains a set of counters $c(n, q, r)$, which for each $q \in \Sigma$ and $r \in \gamma(n)$, records the number of times the response r has been observed when the query q was issued in state $s(n)$. For any edge $(n, q, n') \in E$ and response $r \in \gamma(n')$, if $c(n, q, r) > 0$, it indicates that there exists a sequence from state $s(n)$, transitioning through query q to state $s(n')$ and observing response r.

By analyzing the $\gamma(n)$ sets of each node in the observation tree and the corresponding counters $c(n, q, r)$, the method can quantify the frequency of observing different responses in each state, thereby assessing the uncertainty of the current state machine model and guiding the decisions of the probability model.

In each iteration of active learning, the updates to the observation tree are divided into two steps: recording new observations and detecting conflicts.

1. The steps for recording new observations are as follows: Let the current query sequence be $\pi = \langle q_1, q_2, \ldots, q_n \rangle \in \Sigma^*$, and the actual response sequence from the target device to π be $\rho = \langle r_1, r_2, \ldots, r_n \rangle \in L^*$. Starting from the root node n_r of the observation tree, for $1 \leq i \leq n$, execute the following:

If there exists an edge $(n_{i-1}, q_i, n_i) \in E$, then update $c(n_{i-1}, q_i, r_i) = c(n_{i-1}, q_i, r_i) + 1$; otherwise, create a new node n_i, add it to N, create the edge (n_{i-1}, q_i, n_i), add it to E, and set $\gamma(n_i) = \{r_i\}$ and $c(n_{i-1}, q_i, r_i) = 1$.

Through these steps, the observation tree records the new query-response sequence and updates the corresponding nodes, edges, and counters.

2. The steps for detecting conflicts are as follows: For each query $q_i \in \pi$ in the current query sequence π, execute the following at the observation tree node n_{i-1}: Search the cache tree T_c to see if there exists a node v such that $\text{path}(v) = \text{path}(n_{i-1})$ and $\text{in}(v) = \{q_i\}$.

If such a v exists, compare the actual observed response r_i with $\text{out}(v)$. If $r_i = \text{out}(v)$, it indicates that the response is consistent with the expected output of the current state in the cache tree, and no conflict occurs. If $r_i \neq \text{out}(v)$, it indicates a conflict, meaning that the actual response differs from the expected output when the query q_i is issued in the state corresponding to the current path $\text{path}(n_{i-1})$.

If no such v exists, it also implies a conflict between the current query response and the cache tree. Once a conflict is detected, it is necessary to utilize the counter data from the observation tree and apply the probability model to resolve the conflict and update the cache tree.

The detailed description of the observation tree update process is presented in Algorithm 3.

In the context of learning BLE device state machines, the probability model represents the likelihood that the observed results are inconsistent with the true states due to noise in the wireless environment. For instance, due to signal fading, interference, or device malfunctions, the learner may observe BLE packets or events that do not align with the actual state. Based on the probability model, the learner's decision-making process can be modeled as a Partially Observable Markov Decision Process (POMDP). In the POMDP framework, the true but unobservable set of BLE device states is represented as S, the set of actions the learner can take (sending query commands) is represented as A, and the set of observable output responses is represented as Ω.

Algorithm 3. Observation tree update algorithm

Input: observation tree T_o, query sequence q, response sequence r'
Output: updated observation tree T_o'
 1: **procedure** UPDATEOBSTREE(T_o, q, r')
 2:    RECORDNEWOBSERVATION(T_o, q, r')
 3:    DETECTCONFLICT(T_o, q, r')
 4: **end procedure**
 5: **procedure** RECORDNEWOBSERVATION(T_o, q, r')
 6:    $n \leftarrow root(T_o)$ $\triangleright$ Start from the root node of the observation tree
 7:    **for** $i = 1$ **to** $length(q)$ **do**
 8:        **if** $\exists (n, q[i], n') \in E(T_o)$ **then** $\triangleright$ If there is a corresponding edge
 9:            $c(n, q[i], r'[i]) \leftarrow c(n, q[i], r'[i]) + 1$ $\triangleright$ Update counter
10:            $n \leftarrow n'$ $\triangleright$ Move to the next node
11:        **else** $\triangleright$ Otherwise create a new node and edge
12:            $n' \leftarrow new\_node()$
13:            $N(T_o) \leftarrow N(T_o) \cup \{n'\}$
14:            $E(T_o) \leftarrow E(T_o) \cup \{(n, q[i], n')\}$
15:            $\gamma(n') \leftarrow \{r'[i]\}$ $\triangleright$ Initialize response set
16:            $c(n, q[i], r'[i]) \leftarrow 1$ $\triangleright$ Initialize counter
17:            $n \leftarrow n'$
18:        **end if**
19:    **end for**
20: **end procedure**
21: **procedure** DETECTCONFLICT(T_o, q, r')
22:    $n \leftarrow root(T_o)$ $\triangleright$ Start from the root node of the observation tree
23:    **for** $i = 1$ **to** $length(q)$ **do**
24:        **if** $\neg\exists (n, q[i], n') \in E(T_o)$ **then**
25:            RESOLVECONFLICT($n, q[i], r'[i]$) $\triangleright$ Resolve Conflict
26:        **else if** $r'[i] \notin \gamma(n')$ **then**
27:            RESOLVECONFLICT($n', q[i], r'[i]$) $\triangleright$ Resolve Conflict
28:        **end if**
29:        **if** $\exists (n, q[i], n') \in E(T_o)$ **then**
30:            $n \leftarrow n'$ $\triangleright$ Move to the next observation tree node
31:        **end if**
32:    **end for**
33: **end procedure**

Definition 5. *(Probabilistic Model)*

(1) State transition probability model $P(s'|s, a, \theta)$, which is the probability of transitioning to state s' after executing action a in state s, where θ is the model parameter.

(2) Observation probability model $P(o|s', a, \phi)$, which is the probability of observing o when reaching state s' after executing action a, where ϕ is the model parameter.

In the context of learning BLE device state machines, the action set A corresponds to the input query instruction set I, and the observation set Ω corresponds to the output response set O of the BLE device. The goal of the algorithm is to estimate the parameters θ and ϕ of the state transition probability model $P(s'|s, q, \theta)$ and the observation probability model $P(r|s', q, \phi)$ based on the observed query-response sequences.

Specifically, in the observation tree $O = (N, E, \Sigma, L, \gamma)$, for each node $n \in N$, let $s(n)$ represent the corresponding state, with the state transition probability model defined as:

$$P(s(n')|s(n), q, \theta) = \frac{c(n, q, r)}{\sum_{r' \in \gamma(n)} c(n, q, r')} \tag{2}$$

where the numerator $c(n, q, r)$ represents the count of observing response r when sending query q in the state $s(n)$ at node n. The denominator $\sum_{r' \in \gamma(n)} c(n, q, r')$ is the sum of counts of all possible responses observed after sending query q in the state $s(n)$ at the current node n.

The observation probability model is:

$$P(r|s(n'), q, \phi) = \frac{c(n', q, r)}{\sum_{r' \in O} c(n', q, r')} \tag{3}$$

where the numerator $c(n', q, r)$ is the count of observing response r after node n' sends query q, and the denominator $\sum_{r' \in O} c(n', q, r')$ is the sum of counts of all possible responses observed after node n' sends query q. When a conflict occurs, let the current state be $s(n)$, send query q, and observe a previously unseen response r^*, with the corresponding potential new state being $s(n^*)$. To determine whether to treat this as a new state and add it to the state machine, the algorithm evaluates the following probability:

$$p^* = P(r^*|s(n^*), q, \phi) \times P(s(n^*)|s(n), q, \theta) \tag{4}$$

This considers the observation probability model $P(r^*|s(n^*), q, \phi)$, which represents the probability of observing response r^* when issuing query q in state $s(n^*)$, and the state transition probability model $P(s(n^*)|s(n), q, \theta)$, which represents the probability of transitioning from state $s(n)$ to state $s(n^*)$ after issuing query q. Specifically, the first step is to calculate p^* based on the current count statistics.

$$p^* = \frac{c(n^*, q, r^*)}{\sum_{r' \in O} c(n^*, q, r')} \times \frac{c(n, q, r^*)}{\sum_{r' \in \gamma(n)} c(n, q, r')} \tag{5}$$

The algorithm sets a threshold τ. If $p^* \geq \tau$, the state $s(n^*)$ is considered a new true state and added as a child node of state $s(n)$, while updating the corresponding count, i.e., $c(n, q, r^*) \leftarrow c(n, q, r^*) + 1$. If $p^* < \tau$, then r^* is regarded as a noisy observation, and the state machine is not updated, only the count is updated.

It should be noted that for an observation r^*, relying solely on whether the calculated p^* exceeds the threshold τ is insufficient. It is also necessary to combine this information with subsequent multiple observations to continuously

Algorithm 4. Conflict resolution algorithm

Input: cache tree T_c, observation tree T_o, cache tree node v, response r', threshold τ
Output: updated cache tree T_c, updated observation tree T_o
 1: **procedure** RESOLVECONFLICT(T_c, T_o, v, r')
 2:     $n \leftarrow$ FINDOBSNODE($T_o, path(v)$)   ▷ Find the corresponding node in the observation tree
 3:     **if** $n = $ null **then**
 4:         CREATENEWBRANCH($T_o, path(v), r'$)   ▷ Create a new branch
 5:     **else**
 6:         $p^* \leftarrow$ COMPUTEOBSPROB(n, q, r') $\times$ COMPUTETRANSITIONPROB(n, q, r')
 7:         **if** $p^* \geq \tau$ **then**
 8:             $n^* \leftarrow new\_node()$ ▷ Create a new node to represent a potential new state
 9:             $N(T_o) \leftarrow N(T_o) \cup \{n^*\}$
10:             $E(T_o) \leftarrow E(T_o) \cup \{(n, in(v), n^*)\}$
11:             $\gamma(n^*) \leftarrow \{r'\}$ ▷ Initialize the response set
12:             $c(n, in(v), r') \leftarrow c(n, in(v), r') + 1$ ▷ Update count
13:             UPDATECACHETREE($T_c, path(v), in(v), r'$)   ▷ Update cache tree
14:         **else**
15:             $c(n, in(v), r') \leftarrow c(n, in(v), r') + 1$ ▷ Considered as noise observation, update count
16:         **end if**
17:     **end if**
18: **end procedure**
19: **procedure** COMPUTETRANSITIONPROB(n, q, r')
20:     $numerator \leftarrow c(n, q, r')$
21:     $denominator \leftarrow 0$
22:     **for** $r'' \in \gamma(n)$ **do**
23:         $denominator \leftarrow denominator + c(n, q, r'')$
24:     **end for**
25:     **if** $denominator = 0$ **then**
26:         **return** 0
27:     **else**
28:         $p^* \leftarrow numerator/denominator$
29:         **return** p^*
30:     **end if**
31: **end procedure**

---

track the trend of p^*. If p^* shows a significant increasing trend, there is more justification for considering it as a new state; conversely, if p^* quickly drops close to 0, it can be attributed to noise.

The specific description of the conflict resolution algorithm is provided in Algorithm 4.

4 Experiments

To comprehensively evaluate the performance of the proposed conflict-aware active learning algorithm in learning BLE device state machines, this section designs and conducts a series of experiments. The experiments focus primarily on key aspects such as query efficiency, noise adaptability, parameter sensitivity, and applicability to real devices. By comparing with existing active learning methods, the advantages of the algorithm are validated. Additionally, the experiments explore in depth how factors such as noise, parameter settings, and state machine complexity affect the algorithm's performance, providing a thorough assessment of its robustness and scalability.

4.1 Experimental Setup

To evaluate the performance of the proposed conflict-aware active learning algorithm, two existing algorithms previously applied to active learning of BLE device state machines were selected as benchmarks: the L* algorithm [15] and the TTT algorithm [16].

The experiment utilized a Nordic nRF52840 development board as the BLE central device, where the active learning algorithm was deployed, and an ESP32 development board as the BLE peripheral device, serving as the testing target. To simulate a complex real-world wireless environment, 10 additional BLE devices were deployed within a $20\,m^2$ laboratory. These devices continuously broadcast during operation to generate BLE wireless signal interference. Additionally, two 2.4 GHz WiFi routers and three wireless mice were introduced to create further interference in the 2.4 GHz spectrum, simulating signal disturbances from other wireless devices in real-world conditions.

To assess the algorithm's performance in learning state machines of varying complexities, the experiment designed three BLE state machines with different levels of complexity: a simple protocol state (Simple), simulating basic processes like broadcasting, connection, and data transmission; a medium-complexity protocol state (Medium), which added parameter updates and secure connections on top of the simple protocol state; and a high-complexity protocol state (Complex), which further refined broadcasting behaviors by distinguishing between connectable and non-connectable broadcasts, introduced passive scanning states, and incorporated low-power sleep mode states. By designing these three BLE protocol states of increasing complexity, the experiment aimed to gradually evaluate the algorithm's learning performance across simple, medium, and complex scenarios, providing a comprehensive assessment of its adaptability.

To evaluate the algorithm's performance in different complex environments, the experiment considered three types of noise: random noise, burst noise, and continuous noise. Random noise represents occasional observation errors caused by random interference in the radio environment. Burst noise simulates continuous observation failures over a period of time, such as temporary device disconnections. Continuous noise is attributed to persistent errors caused by firmware defects or hardware malfunctions. Random noise was introduced by randomly replacing each state observation value with another state value based on a prede-fined probability in the monitoring program that observes the state interactions between the two BLE devices. Burst noise was simulated by temporarily discon-necting the target BLE device during the learning process, mimicking sudden connection loss. Continuous noise was generated by activating a debug switch in the BLE device firmware, forcing the device into an abnormal state that pre-vents it from transitioning correctly to normal BLE protocol states. The noise intensity was categorized into three levels: low, medium, and high, corresponding to observation error rates of 5%, 10%, and 20%, respectively.

To quantitatively evaluate the algorithm's performance, the study adopted the following three metrics:

(1) Number of Queries: This refers to the total number of queries required for the algorithm to successfully learn the state machine. Fewer queries indicate higher query efficiency, meaning the algorithm can construct the correct device state machine more quickly.

(2) Number of States: This refers to the total number of states included in the final state machine. A higher number of states suggests a more detailed modeling of the target.

(3) Convergence Time: This is the time taken from the start of the algo-rithm to convergence, meaning no further state updates occur. Shorter conver-gence times indicate faster convergence, allowing the algorithm to reach the final device state machine in a shorter period.

4.2   Experimental Results

In an ideal environment without any external interference, the experiment first evaluated the baseline performance of the conflict-aware active learning algo-rithm. As shown in Fig. 2, the algorithm achieved the fewest number of queries, the highest number of discovered states, and the shortest convergence time under interference-free conditions. This result provides a benchmark for subsequent tests involving the introduction of interference and adjustments to the algo-rithm's parameters.

To assess the impact of noise on the algorithm's performance, the experi-ment introduced three types of noise (random noise, burst noise, and continuous noise) at low, medium, and high intensity levels. As shown in Fig. 3, different types and intensities of noise reduced the algorithm's query efficiency, state dis-covery capability, and convergence speed to varying degrees. Continuous noise had the greatest impact on the algorithm, while random noise had the least. When the noise intensity was high, the performance degradation across all met-rics was most significant. However, even under the most adverse conditions of high-intensity continuous noise, the algorithm demonstrated strong robustness.

The experiment also evaluated the impact of algorithm parameters on per-formance, focusing on a key parameter, τ, in the conflict-aware algorithm. The parameter τ controls the threshold for updating the probabilistic model, deter-mining the degree to which the algorithm trusts new observations and updates the model. The experiment tested five different values of τ under 0.1, 0.3, 0.5,

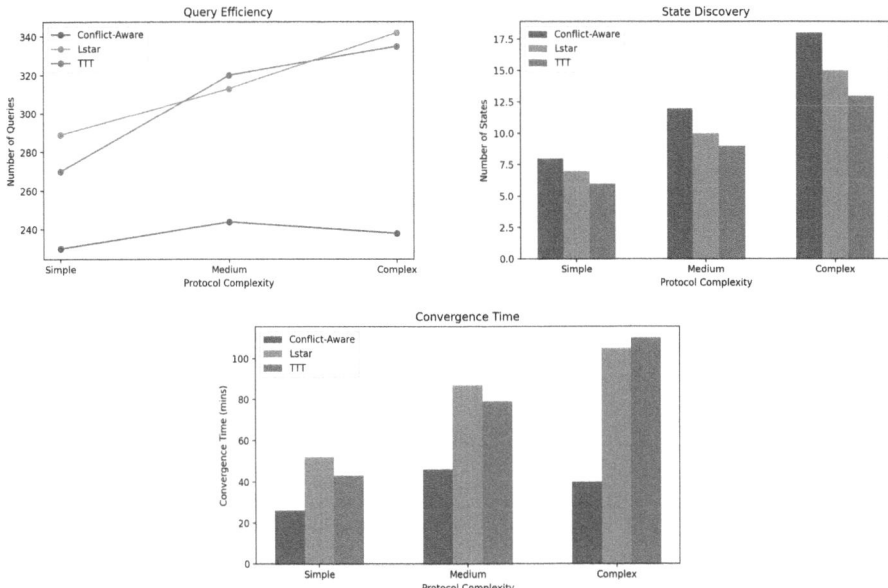

Fig. 2. The conflict-aware active automata learning algorithm baseline.

0.7, and 0.9 to assess the algorithm's performance in terms of the number of queries and convergence time. The results are shown in Fig. 4.

Table 1 presents the L2CAP state coverage for various BLE fuzzing schemes. The complete BLE state machine consists of 18 states. Compared to other BLE fuzzers, our proposed fuzzer detects the most states of the BLE state machine at a general noise intensity. Furthermore, we extend the evaluation to test the state coverage ratios across different BLE devices (i.e., ESP32, CC2640R2, nRF52832, and Zephyr). The experimental results, summarized in Table 2, demonstrate that our fuzzer outperforms three BLE fuzzing benchmarks: Random Mutation Fuzzing (RMF), State Machine Model-based Fuzzing (SMMF), Heuristic Rule-based Mutation Fuzzing (HRMF), achieving the highest state coverage ratios across all four BLE devices. These results validate the effectiveness of our proposed BLE state machine construction method.

Table 1. L2CAP state coverage by different fuzzers

	This work	L* [15]	TTT [16]	Defensics [12]	BFuzz [17]	BSS [18]
Covered states	17	12	13	7	6	3

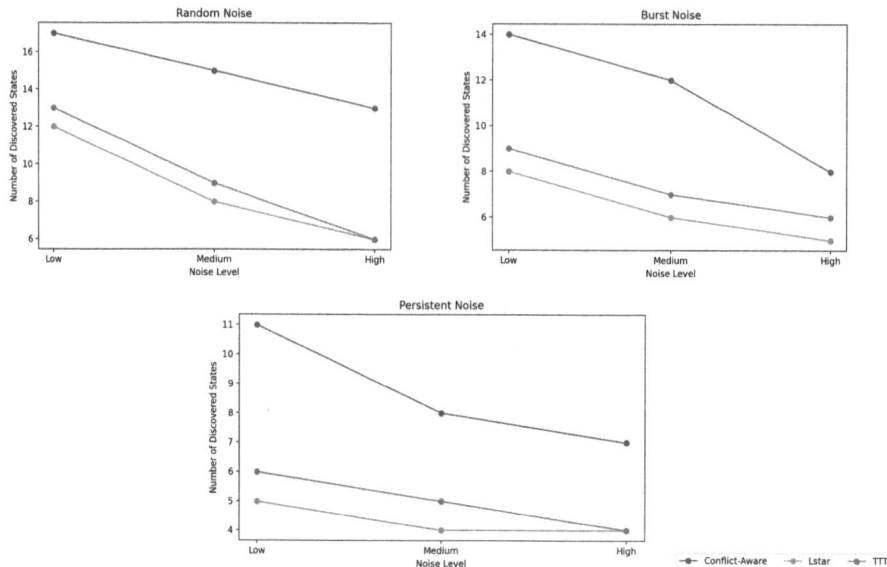

Fig. 3. Performance of algorithm in noisy environments.

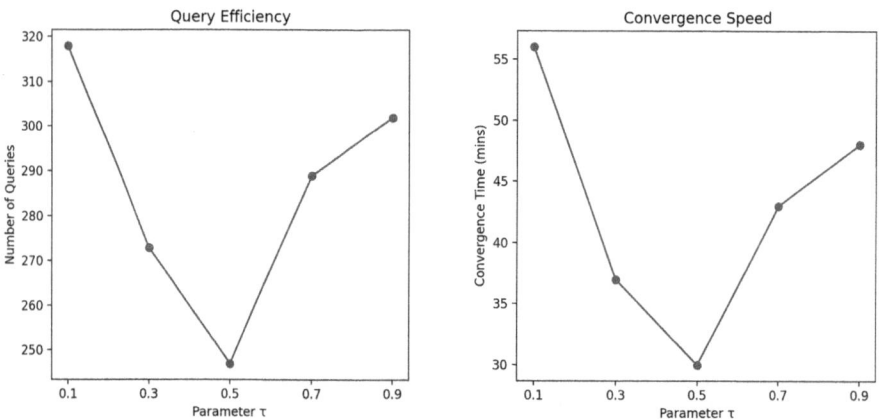

Fig. 4. Effect of algorithm parameter τ on performance.

Table 2. State Coverage ratios in different BLE devices

	This work	RMF	SMMF [15, 16]	HRMF [19]
ESP32	88.9%	22.2%	61.1%	72.2%
CC2640R2	83.3%	16.7%	55.5%	66.7%
nRF52832	77.8%	11.1%	50%	61.1%
Zephyr	94.4%	22.2%	66.7%	77.8%

5   Conclusion

This paper proposes a conflict-aware active learning method for constructing BLE device state machine models in complex wireless environments. The method integrates components such as a cache tree, distinguishing tree, observation tree, and probabilistic model to effectively address the inefficiencies of existing active learning algorithms in generating BLE device state machines, as well as to overcome the limitations when dealing with uncertain behaviors of BLE devices.

Specifically, the cache tree is used to initialize the BLE protocol state machine, store, and reuse previous query sequences, significantly reducing the number of queries. The distinguishing tree guides the generation of optimal distinguishing query sequences, improving query efficiency. The observation tree and probabilistic model work together to track and quantify uncertainty during the learning process. When conflicts arise, the probabilistic model calculates the likelihood of various responses, guiding the algorithm to make informed decisions and dynamically update the state machine model.

The method's performance in terms of query efficiency, noise adaptability, and parameter sensitivity was thoroughly evaluated through experiments. The results confirm that the proposed method offers significant advantages over existing active learning methods. It effectively handles various interference noises in complex wireless environments, accurately and efficiently constructing BLE device state machines. This method provides a powerful tool for the automated modeling of BLE devices and lays a solid foundation for BLE device fuzz testing.

Funding Information. This work was supported in part by the National Natural Science Foundation of China under grants 62372096.

References

1. Seri, B., Vishnepolsky, G., Zusman, D.: BLEEDINGBIT: the hidden attack surface within BLE chips. Technical report (2019)
2. Fiterau-Brostean, P., Jonsson, B., Merget, R., De Ruiter, J., Sagonas, K., Somorovsky, J.: Analysis of DTLS implementations using protocol state fuzzing. In: 29th USENIX Security Symposium (USENIX Security 20), pp. 2523–2540 (2020)
3. Aichernig, B.K., Muškardin, E., Pferscher, A.: Learning-based fuzzing of IoT message brokers. In: 2021 14th IEEE Conference on Software Testing, Verification and Validation (ICST), pp. 47–58. IEEE (2021)
4. Zou, Y.H., Bai, J.J., Zhou, J., Tan, J., Qin, C., Hu, S.M.: TCP-fuzz: detecting memory and semantic bugs in TCP stacks with fuzzing. In: 2021 USENIX Annual Technical Conference (USENIX ATC 21), pp. 489–502 (2021)
5. Wu, J., et al.: BLESA: spoofing attacks against reconnections in Bluetooth low energy. In: 14th USENIX Workshop on Offensive Technologies (WOOT 20) (2020)
6. Wu, J., Wu, R., Xu, D., Tian, D.J., Bianchi, A.: Formal model-driven discovery of Bluetooth protocol design vulnerabilities. In: 2022 IEEE Symposium on Security and Privacy (SP), pp. 2285–2303. IEEE (2022)
7. Zhao, B., et al.: StateFuzz: system call-basedstate-aware linux driver fuzzing. In: 31st USENIX Security Symposium (USENIX Security 22), pp. 3273–3289 (2022)

8. Zalewski, M. AFL(american fuzzy lop). https://github.com/google/AFL. Accessed 25 July 2019
9. Fuzzing, I.W.: SAGE: Whitebox Fuzzing for Security Testing, vol. 10, no. 1. SAGE, Upper Saddle River (2012)
10. Garbelini, M.E., Bedi, V., Chattopadhyay, S., Sun, S., Kurniawan, E.: BrakTooth: causing havoc on Bluetooth link manager via directed fuzzing. In: 31st USENIX Security Symposium (USENIX Security 22), pp. 1025–1042 (2022)
11. Garbelini, M.E., Wang, C., Chattopadhyay, S., Sumei, S., Kurniawan, E.: Sweyn-Tooth: unleashing mayhem over Bluetooth low energy. In: 2020 USENIX Annual Technical Conference (USENIX ATC 20), pp. 911–925 (2020)
12. Synopsys: Defensics Fuzz Testing. https://www.synopsys.com/software-integrity/security-testing/fuzz-testing.html
13. Ruge, J., Classen, J., Gringoli, F., Hollick, M.: Frankenstein: advanced wireless fuzzing to exploit new Bluetooth escalation targets. : 29th USENIX Security Symposium (USENIX Security 20), pp. 19–36 (2020)
14. Park, H., Nkuba, C.K., Woo, S., Lee, H.: L2Fuzz: discovering Bluetooth L2CAP vulnerabilities using stateful fuzz testing. In: 2022 52nd Annual IEEE/IFIP International Conference on Dependable Systems and Networks (DSN), pp. 343–354. IEEE (2022)
15. Pferscher, A., Aichernig, B.K.: Stateful black-box fuzzing of Bluetooth devices using automata learning. In: NASA Formal Methods Symposium, pp. 373–392. Springer, Cham (2022)
16. Karim, I., Al Ishtiaq, A., Hussain, S.R., Bertino, E.: Blediff: scalable and property-agnostic noncompliance checking for BLE implementations. In: 2023 IEEE Symposium on Security and Privacy (SP), pp. 3209–3227. IEEE (2023)
17. Kim, S., Woo, S., Lee, H., Oh, H.: Poster: IoTcube: an automated analysis platform for finding security vulnerabilities. In: Proceedings of the 38th IEEE Symposium on Poster presented at Security and Privacy (2017)
18. Betouin, P.: [Infratech - release] version 0.6 de Bluetooth Stack Smasher (2006). http://www.secuobs.com/news/05022006-bluetooth10.shtml
19. Ma, R., Wang, D., Hu, C., Ji, W., Xue. J.: Test data generation for stateful network protocol fuzzing using a rule-based state machine. In: Tsinghua Science and Technology, vol. 21, no. 3, pp. 352–360 (2016). https://doi.org/10.1109/TST.2016.7488746
20. Chen, J., et al.: Discovering memory corruptions in IoT through app-based fuzzing, IoTFuzzer. In: NDSS (2018)
21. Feng, X., et al.: Snipuzz: black-box fuzzing of IoT firmware via message snippet inference. In: Proceedings of the 2021 ACM SIGSAC Conference on Computer and Communications Security, pp. 337–350 (2021)

Optimizing Indoor Network Element Layout for Enhanced Signal Coverage and Security in Location-Based Services

Xiaomin Yu[1,2]($\boxtimes$) and Xiaokun Yu[3]($\boxtimes$)

[1] College of Computer and Control Engineering, Qiqihar University, Qiqihar, China
yuxiaomin@qqhru.edu.cn
[2] Heilongjiang Key Laboratory of Big Data Network Security Detection and Analysis, Qiqihar University, Qiqihar, China
[3] Heilongjiang Jiaotong Polytechnic, Qiqihar, China
309569972@qq.com

Abstract. Recently, with the rapid development of communication technology, location-based services (LBS) in indoor environments have penetrated into various aspects of people's daily lives. Wireless local area networks (WLAN) have become the preferred choice for location-based services due to their broad deployment and low cost. However, the typical non-line-of-sight characteristics of indoor environments result in severe fluctuations in received signal strength (RSS), signal coverage vulnerabilities can exacerbate environmental interference, making positioning results more vulnerable to attacks or tampering. The layout of indoor network elements is a key factor that constrains signal coverage rate. This paper focuses on the optimization of indoor network element layout, proposes an indoor network element layout optimization model based on signal coverage and quality in indoor environments and introduces an Adaptive Simulation Annealing Non-inertial Opposite Particle Swarm Optimization algorithm (ASA-NRPSO). Firstly, a reverse strategy is introduced into the particle flight process of the Particle Swarm Optimization (PSO) algorithm, and high-quality particles are selected for the next generation iteration. In the selection of the reverse strategy, the simulated annealing idea is integrated to adaptively select the reverse strategy based on the needs of the particles at different stages, which avoids the particles falling into local optima. The inertia term in the particle swarm velocity update formula is replaced with group information for non-inertial updates. This can more fully utilize the global information of the population to guide the movement of the next generation of particles and improve the convergence speed of the particle swarm. The elite mutation strategy is used to increase the diversity of particles and improve the global search capability of the algorithm. This method ensures the rapid generation of network element layout to improve the signal coverage rate and positioning accuracy.

Keywords: Indoor positioning · Network element layout · Adaptive · Optimization model · Security

© The Author(s), under exclusive license to Springer Nature Switzerland AG 2025
W. Meng et al. (Eds.): ADIoT 2024, LNCS 15397, pp. 53–75, 2025.
https://doi.org/10.1007/978-3-031-85593-1_4

1   Introduction

1.1   Motivation

With the rapid advancement of mobile computing and wireless communication technologies, the Internet of Things (IoT) enhances the mobile computing landscape by interconnecting billions of intelligent terminals with various devices. There are increasingly more location-based services (LBS) applied to smart terminals. Location information can not only be used for map and navigation applications, but also for social networks, public safety, and military purposes. This has led to a continuous increase in people's demand and expectations for LBS. Positioning technology will have a significant impact on the performance, reliability, and privacy of LBS systems. The WLAN indoor positioning technology based on RSS has attracted the attention of researchers due to its advantages of high positioning accuracy, fast positioning speed, and low deployment cost. A key issue affecting indoor positioning accuracy is signal coverage and signal quality. Due to the initial indoor network element layout being designed for communication purposes, it cannot meet the requirements of positioning. Need to optimize and adjust the original network element layout. Effectively utilize existing infrastructure while minimizing the deployment of new network elements while ensuring positioning accuracy. Due to the non line of sight (NLOS) characteristics of indoor environments, signals undergo reflection, refraction, and diffraction during propagation, resulting in loss and even the inability of positioning terminals to be covered by multiple network elements simultaneously, leading to increased positioning errors. RSS is susceptible to external environmental interference, and signal coverage vulnerabilities can exacerbate environmental interference, making positioning results more vulnerable to attacks or tampering. Attackers may mislead the positioning system by manipulating signal strength or simulating signal sources, thereby disrupting the positioning process and affecting the security of the entire network. A reasonable network element layout will not only improve signal coverage, but also increase the number of first path signals received by terminals, ensuring positioning stability and improving positioning accuracy.

Currently, large public places such as schools, factories, and hospitals etc. have installed wireless network access points (AP), ensuring that users can receive signals from at least one AP and communicate any position. But, for indoor positioning, each location must receive a sufficient number of reference signals to achieve positioning. That is, the coverage of communication network elements is single coverage, and the coverage of positioning network elements is K coverage (K is not less than 4). Therefore, the standard for network element layout for high-precision positioning is different from traditional network element layout which only considers signal accessibility. At the same time, the structure of the indoor environment, wall structure, decoration materials, various furniture, decorations, etc. will interfere with wireless signal propagation. Weaker signals are greatly affected by noise interference, resulting in larger fluctuations and decreased reference signal quality. Signal quality and coverage largely depend on the relative geometric shape between the positioning point and the network element [1]. Therefore, this paper mainly researches improving positioning accuracy through optimizing network element layout.

During indoor positioning, it was found that the locations with larger errors are those that cannot receive a sufficient number of AP signals or have weak AP signal reception, which means that they are not covered by the positioning signal. In these locations, even the most advanced positioning algorithms cannot be effective. Therefore, signal coverage and signal quality are the basis and guarantee for achieving high-precision positioning, and reasonable network element layout is the key to ensuring signal coverage and improving signal quality. The optimization of network element layout mainly includes the two aspects of signal quality and coverage. Firstly, when placing network elements, each point in the area should be covered by at least four network elements. Secondly, the geometric dilution of precision (GDOP) should be minimized to reduce the average positioning error. Network element layout optimization is the process of selecting multiple locations suitable for placing network elements from several positions within the area to achieve full signal coverage and minimize positioning errors. Currently, commonly used methods mainly include station layout based on geometric shape and optimization deployment based on swarm intelligence algorithms. Geometric layout can achieve simple communication signal coverage, but due to the complexity of indoor scenes, the application of this method is greatly restricted. It is difficult to find a deterministic general solution for the network element layout problem through the combination of geometric layouts, and exhaustive search for the optimal solution is also impractical. Therefore, we can only seek suboptimal solutions as the solution to this problem. The method based on swarm intelligence optimization is widely used, but single intelligent algorithms have their inherent shortcomings. The fusion of multiple swarm intelligence algorithms can complement the shortcomings of a single algorithm.

1.2   Main Contributions

Due to the NP-hard nature of indoor network element layout, it is difficult to find the optimal solution. Therefore, this article proposes an indoor network element layout optimization model. On the basis of this model, the Adaptive Simulated Annealing Non-Inertial Reverse Particle Swarm Optimization (ASA-NRPSO) algorithm was proposed to optimize the layout of network elements. Firstly, the idea of simulated annealing is integrated into the traditional particle swarm optimization algorithm. Based on the needs of particles at different stages, adaptive selection of reverse strategies is carried out to optimize the particle swarm and avoid particles falling into local optima; Secondly, the inertia free particle swarm updating formula was introduced to more fully utilize population information to guide the direction of motion of the next generation of particles, further enhancing the local development ability of the population and improving the convergence speed of the particle swarm; Once again, an adaptive elite mutation strategy is adopted to improve the global search capability of the algorithm, in order to generate the optimal network element layout. By optimizing the layout of network elements, we can effectively improve the coverage of positioning signals, ensure positioning stability, and reduce the interference of multipath and non line of sight noise on signals during the measurement process, improve signal quality, and ultimately achieve the goal of overall error suppression and improved positioning accuracy.

The rest of the paper is organized as follows. Section 2 reviews related works on network element layout optimization and indoor positioning systems. In Sect. 3, we present our proposed adaptive network element layout optimization method for high-precision positioning, which includes the introduction of a novel Particle Swarm Reverse Fusion Optimization Algorithm. Section 4 details the experimental setup and analysis, where we validate the effectiveness of our proposed method through a series of experiments conducted in real-world scenarios. The conclusions and implications of our findings are discussed in Sect. 5.

2   Related Works

Indoor wireless positioning has always been a popular research topic because it is the foundation of location-based services. WLAN based on IEEE 802.11 is widely used in many indoor environments, providing data communication for mobile devices. Its high bandwidth, free spectrum, and reliability give users a good data communication experience [16,19]. In recent years, thousands of APs have been rapidly deployed in many public places such as university campuses, airports, and train stations. Recent studies have shown that APs deployed in WLANs are rarely used to their peak capacity, and most APs are often idle [5,15]. These WLAN systems were not originally designed for positioning services, so the accuracy of positioning cannot be guaranteed. To meet the positioning requirements, it is necessary to optimize the layout of WLAN access points to improve the coverage of positioning signals and the accuracy of positioning [11]. The purpose of network element layout optimization is to solve two problems. On the one hand, effective improvement of the coverage of positioning signals through network element optimization layout can ensure positioning stability. On the other hand, by optimizing the layout of network elements, the noise generated by multipath and non-line-of-sight in the measurement process can be reduced, the signal quality can be improved, and the overall error suppression and positioning accuracy can be improved.

In some applications, when the network is dense enough, area coverage can be approximated by ensuring point coverage. In this case, all points of wireless devices can be used to represent the entire area. Many applications related to safety and reliability require always ensuring K-coverage in the area [3]. Reference [3] uses linear programming to derive two non-global solutions for K-coverage. For K-coverage, a global method is proposed in reference [9], which constructs K independent sets by constructing K control sets through K connections, each set implementing one coverage and these sets together provide K-coverage. Reference [4] provides a positioning solution for solving the same problem. When covering a single target, the dominant set algorithm for implementing point coverage is considered [14].

In indoor positioning, the deployment of indoor network elements is a key issue to ensure positioning stability while improving positioning accuracy. The selection of network element locations has always been regarded as an NP-hard problem [18,26]. Even if only a large-grain representation of the search space is used, it is impossible to perform search through enumeration due to the multiple combinations of layouts [2]. Therefore, the solution to the network element layout can only be a suboptimal solution or an approximate solution [24]. Currently, optimization methods for network element layout include random geometry and heuristic search, the latter of which can accelerate the search speed.

In the field of random geometry, reference [10] studied the necessary observation space to achieve the minimum Geometric Dilution of Precision (GDOP) for absolute range measurement in two-dimensional wireless positioning systems. The Angle of Coverage (AOC) was introduced as a parameter to describe the observation space, reflecting the geometric relationship between the target and measurement points. A new intelligent geometric relationship between the target and N measurement points was proposed. Then, it was proven that this geometric relationship has the minimum AOC as the necessary AOC in a two-dimensional wireless positioning system to achieve the minimum GDOP. When the AOC is less than the necessary AOC, the optimal geometric shape and its new GDOP lower limit will be obtained. Reference [12] modeled the layout of cellular networks using homogeneous Poisson distribution, assuming complete independence among base stations. Bais et al. designed the indoor base station layout as a square [2], mainly considering solving the signal coverage problem and reducing positioning error by improving signal coverage rate. However, due to the irregularity of buildings, it is difficult for all base stations to have a square layout. Zhou et al. improved this method by placing four base stations in a rectangular shape and evaluating the positioning performance and effect [28], confirming the theory proposed by Chen et al. that placing four base stations in a rectangular shape can achieve the best positioning accuracy [6].

Regarding heuristic search, intelligent algorithms are more commonly used in network element layout optimization due to their strong adaptability and ease of modeling [7]. Chen et al. proposed to combine adaptive genetic algorithm with ant colony algorithm to optimize base station layout. The average positioning error was significantly improved, but the algorithm convergence speed did not show significant improvement, and the integration of the two algorithms was not achieved [7]. Maneerat et al. proposed a simulated annealing-based network element layout optimization algorithm, but this method has poor global search performance, poor algorithm stability, and is highly susceptible to the initial parameter settings and initial layout of network elements [17]. Wang et al. used genetic algorithm to correct positioning errors and improve positioning accuracy. But, in the iterative optimization process, due to the inherent defects of genetic algorithm, it is susceptible to falling into local optima, failing to obtain the global optimal solution [21,22]. Zhou et al. obtained the parameters of the wireless propagation model based on the data collected from existing access points. Based on self-learning parameters, they proposed a genetic algorithm-based AP layout optimization method [27]. From the above literature analysis, it can be seen that any single algorithm has certain performance defects and cannot achieve satisfactory results. Therefore, domestic and foreign scholars have turned their research to the integration optimization of multiple intelligent algorithms [20,23].

Gharghan et al. proposed a Particle Swarm Optimization and Artificial Neural Network (PSO-ANN) fusion algorithm. The algorithm uses a feedforward neural network model and uses the Levenberg-Marquardt training algorithm to estimate the distance between mobile nodes and anchor nodes. The positioning accuracy of the algorithm has been improved, but feedforward neural network training requires a large number of samples, otherwise it cannot converge to the global minimum or a sufficiently good local minimum value [8].

Kang et al. proposed a hybrid simulated annealing genetic optimization algorithm to improve the convergence speed of genetic algorithm. In the early stage, standard genetic algorithm is used for optimization, and annealing operation is performed for the optimization results of genetic algorithm. The algorithm improves the positioning accuracy, but it is easy to introduce random values in the later stage, making the algorithm unable to converge to the extremum, and the algorithm stability is poor [13].

From the above literature, it can be seen that research on network element layout optimization mainly focuses on signal coverage model research and optimization algorithm research. K-fold coverage is commonly used for signal coverage. There are mainly two methods for network element layout optimization: random geometry and heuristic search. Random geometry method is suitable for outdoor environments, while heuristic search algorithms are suitable for indoor environments. As a single heuristic search algorithm has its own performance defects, complementary fusion algorithms are currently a research hotspot. When integrating algorithms, positioning accuracy, convergence speed and stability should be considered. Existing network element layout optimization methods have improved to a certain extent in terms of positioning accuracy and signal coverage rate, but there is still further improvement in algorithm fusion.

3   Adaptive Network Element Layout Optimization Method for High-Precision Positioning

Since the network element layout problem is an NP-Hard problem, the optimal solution cannot be obtained through exhaustive methods. We can only seek suboptimal solutions through intelligent algorithms. Different intelligent algorithms have their own advantages and disadvantages, so algorithm fusion is adopted in network element layout optimization. In swarm intelligence algorithms, particle swarm optimization has the advantages of simplicity, fast convergence speed, and few parameter settings, but the disadvantage is that it is easy to fall into local optima and premature convergence. Simulated annealing algorithm has strong local search ability, which can make particles jump out of the current area and enter another search area. However, its optimization performance in the entire search space is poor and it cannot guarantee that it will enter the optimal search area. The moderate fusion of the two algorithms, using their respective advantages, will inevitably improve the convergence speed and solution quality of the algorithm.

In this paper, a reverse strategy is introduced in the particle swarm optimization algorithm to improve the quality of particles. On this basis, an elite mutation strategy is adopted to enhance the diversity of particles and avoid local optima. In the selection of reverse strategy, simulated annealing mechanism is introduced to adaptively select the reverse strategy according to the state of particles. This truly integrates simulated annealing and particle swarm optimization, and then improves the convergence speed and optimization quality of the algorithm.

3.1   Network Element Combination Layout Parameter Design

Table 1. Point properties description

Attribute Name	Meaning	Type
x	X-axis coordinate	float
y	Y-axis coordinate	float
z	Z-axis coordinate	float

Table 2. Net properties description

Attribute Name	Meaning	Type
id	Network Element ID	int, must be non-negative
netloc	Network Element Coordinate	Point
r	Network Element Coverage Radius	float

Before optimizing the layout of network elements, spatial division is first performed according to the network element coverage model based on the specific positioning scenario. The positions of network elements and positioning points are both represented by the center of small cubes. The properties of the center point-Point are shown in Table 1. Each center point-Point is represented by the three-dimensional coordinates of the X-axis, Y-axis, and Z-axis, all of which are float type data.

The properties of the network element object Net are described in Table 2. The id is a non-negative integer number for each network element. The netloc represents the true coordinates of the network element, which is three-dimensional coordinate data of the Point type. The r is the coverage radius of the network element.

The properties of each terminal waiting to be positioned in the space are described in Table 3. The id is a non-negative integer number for the terminal. The actual coordinates real-loc and estimated coordinates est-loc are both represented by three-dimensional coordinate data of the Point type. The error is the positioning error, represented by the Euclidean distance between the actual coordinates and the true coordinates, The k is the number of network elements covering the terminal.

Table 3. Terminal properties description

Attribute Name	Meaning	Type
id	Terminal ID	int, must be non-negative
est-loc	Terminal estimated coordinates	Point
real-loc	Terminal actual coordinates	Point
error	positioning error	float
k	Number of network elements covering the terminal	int

Table 4. Layout properties description

Attribute Name	Meaning	Type
id	network layout ID	int, must be non-negative
netnum	network layout numbers	int
scheme	network layout list	Net[]
error	network layout fitness	float
cover	coverage ratio	float

Table 4 lists the properties of the Layout object, which represents the layout of network elements. The id is the ID number of the network layout, a non-negative integer. The netnum is the number of network elements included in the layout. The scheme is the list of network elements in the current layout. The error is the fitness of the current layout, represented by the mean positioning error. The cover is the spatial signal coverage ratio of the layout.

3.2 Particle Swarm Reverse Fusion Optimization Algorithm

Particle Swarm Optimization (PSO) algorithm is a stochastic biomimetic evolutionary algorithm proposed during the study of group foraging behavior such as fish, ants, and birds. It has been successfully applied in various fields such as intelligent transportation, intelligent manufacturing, biotechnology, and national defense. However, the PSO algorithm has the drawbacks of excessive parameter dependence and easy local optimization. To address these issues, this paper improves the traditional PSO algorithm to overcome its shortcomings.

Standard PSO Algorithm. The PSO algorithm is a stochastic evolutionary algorithm for swarm intelligence. Each particle in the swarm represents a candidate solution to the problem, and in the solution space, the particle searches for the optimal solution according to the dynamic equations of formula (1) and (2).

$$v_{ij}(t+1) = \varpi v_{ij}(t) + c_1 rand\left(pbest_{ij}(t) - x_{ij}(t)\right)$$
$$+ c_2 rand\left(gbest_j(t) - x_{ij}(t)\right) \tag{1}$$

$$x_{ij}(t+1) = x_{ij}(t) + v_{ij}(t+1) \tag{2}$$

Among them, $v_{ij}(t)$ and $x_{ij}(t)$ represent the velocity and position of the i-th particle at the j-th dimension respectively; $pbest_{ij}(t)$ and $gbest_j(t)$ represent the best position and global best position of the i-th particle at the j-th dimension in the t-th iteration. $i = 1, 2, \ldots, N, j = 1, 2, \ldots, D$ (where N is the swarm size and D is the dimension of the search space). $\varpi \in [0, 1]$ represents the inertia weight. $c_1, c_2 \in [0, 2]$ represent the cognitive learning factor and social learning factor respectively, $c_1 = c_2$ generally, and rand is a uniformly distributed random number between 0 and 1.

Reverse PSO Algorithm. The main idea of opposition-based learning (OBL) is to simultaneously evaluate a feasible solution and its opposite solution, and select the better one for the next generation evolution. The general definition of OBL is as follows:

Let $x_i = (x_{i1}, x_{i2}, \ldots, x_{iD})$ be a particle in the D-dimensional space, and its opposite solution $ox_i = (ox_{i1}, ox_{i2}, \ldots, ox_{iD})$ can be obtained by Eq. (3).

$$ox_i = k(da + db) - x_i \tag{3}$$

Among them, $k \in [0, 1]$ is a random number, da and db are the left and right boundaries of the search space, and each dimension takes the following values:

$$da = \min\left(x_{ij}\right), db = \max\left(x_{ij}\right) \tag{4}$$

If the reverse solution ox_i exceeds the boundary, a random solution within the interval $[da, db]$ will be generated for reset.

$$ox_i = rand(da, db), ox_i < da \text{ or } ox_i < db \tag{5}$$

From Eq. (1), it can be seen that the adjustment of the search direction for each particle in the search space depends on the dynamic equation, which consists of two characteristics: the velocity that describes the particle's motion and the position that describes the particle's stillness. The velocity formula consists of three parts: the inertia part that is affected by personal motion information, the cognition part that is affected by particle memory information, and the social part that is affected by group information. The first two parts belong to the self-correlated information part, affected by the current particle motion and memory, while the third part belongs to the overall correlated information part, affected by the entire group. Obviously, group information is rarely used to guide particle motion. In addition, the flight direction of the particle is greatly influenced by inertia, which reflects the particle's trust in various prior knowledge. However, although the inertia term can bring diversity to the group, unreasonably setting the inertia parameter in the velocity update formula may reduce the convergence speed of PSO.

Inspired by human social behavior, where good experience exchange and cooperation among individuals in a team are more conducive to achieving overall goals, a new velocity update formula was designed to replace the traditional PSO [13]. The modified velocity formula does not include the inertia component, i.e., NIV (No Inertial component Velocity). The dynamic equation is defined in (6) as follows:

$$
\begin{aligned}
v_i(t+1) = s \cdot (U(t) - U(t-1)) &+ c_1 \operatorname{rand}(pbest_i(t) - x_i(t)) \\
&+ c_2 \operatorname{rand}(gbest(t) - x_i(t))
\end{aligned}
\tag{6}
$$

The variables $U(t)$ and $U(t-1)$ are the average group positions of the t-th and (t-1)-th generations respectively. The s represents differential coefficient of the search range. Formula (6) can more fully utilize the group information to guide the movement direction of the next generation of particles, which converges faster, but the possibility of falling into local optima still exists.

Theoretical Analysis. Next, we will analyze the convergence of formula (6) theoretically and determine the stable region of the particle swarm system. Firstly, we determine the movement model of the particle swarm.

Substituting Eq. (6) into Eq. (2), we get:

$$
\begin{aligned}
x_{ij}(t+1) &= x_{ij}(t) + s \cdot (U_j(t) - U_j(t-1)) + c_1 \cdot \operatorname{rand}(pbest_{ij}(t) - x_{ij}(t)) \\
&+ c_2 \cdot \operatorname{rand}(gbest_j(t) - x_{ij}(t)) \\
&= x_{ij}(t) \cdot (1 - c_1 \cdot \operatorname{rand} - c_2 \cdot \operatorname{rand}) + s \cdot (U_j(t) - U_j(t-1)) \\
&+ c_1 \cdot \operatorname{rand} \cdot pbest_{ij}(t) + c_2 \cdot \operatorname{rand} \cdot gbest_j(t)
\end{aligned}
\tag{7}
$$

Let $\varphi_1 = c_1 \cdot \operatorname{rand}, \varphi_2 = c_2 \cdot \operatorname{rand}, \varphi = \varphi_1 + \varphi_2$. Thus, (7) can be further simplified to:

$$
\begin{aligned}
x_{ij}(t+1) = x_{ij}(t) \cdot (1 - \varphi) &+ s \cdot (U_j(t) - U_j(t-1)) \\
&+ \varphi_1 \cdot pbest_{ij}(t) + \varphi_2 \cdot gbest_j(t)
\end{aligned}
\tag{8}
$$

From (8), we can obtain the iterative formula for the (t+2)-th generation:

$$x_{ij}(t+2) = x_{ij}(t+1) \cdot (1-\varphi) + s \cdot (U_j(t+1) - U_j(t))$$
$$+ \varphi_1 \cdot pbest_{ij}(t+1) + \varphi_2 \cdot gbest_j(t+1) \tag{9}$$

Let $O = s \cdot (U_j(t+1) - U_j(t)) + \varphi_1 \cdot pbest_{ij}(t+1) + \varphi_2 \cdot gbest_j(t+1)$, The O Known as the stable point, the second-order difference equation for particle motion can be expressed as:

$$x_{ij}(t+2) = x_{ij}(t+1) \cdot (1-\varphi) + O_i \tag{10}$$

The matrix representation corresponding to formula (10) is:

$$\begin{bmatrix} x^{t+2} \\ x^{t+1} \\ 1 \end{bmatrix} = A \cdot \begin{bmatrix} x^{t+1} \\ x^t \\ 1 \end{bmatrix} \tag{11}$$

Here $A = \begin{bmatrix} 1-\varphi & 0 & 0 \\ 1 & 0 & 0 \\ 0 & 0 & 1 \end{bmatrix}$, the characteristic equation corresponding to the coefficient matrix A is:

$$|A - I\lambda| = \lambda(1 - \varphi - \lambda)(\lambda - 1) = 0 \tag{12}$$

Obtain $\lambda_1 = 0, \lambda_2 = 1 - \lambda, \lambda_3 = 1$ The trajectory of the particle's movement is as follows:

$$x(t) = k_1 + k_2 \cdot \lambda_2(t) \tag{13}$$

Both k_1 and k_2 are constants, which determined by the initial conditions of the dynamic system. Substituting (13) into (11), obtain $k_1 = \frac{1}{\varphi} \cdot O, k_2 = x(0) + \frac{1}{\varphi} \cdot O$, where $k_1 = \frac{1}{\varphi} \cdot O, k_2 = x(0) + \frac{1}{\varphi} \cdot O$ is the initial position of the particle. If the particle's trajectory converges, formula (13) needs to satisfy $\varphi > 1$ and $\|\lambda_2\| < 1$, that is, $\|1 - \varphi\| < 1$, which leads to $0 < \varphi < 2$. Therefore, the second-order stable region can be obtained as follows:

$$SU_D = \{\varphi : 1 < \varphi < 2\} \tag{14}$$

In SU_D, the trajectory sequence of particles will converge stably to the stable point O.

$$\lim_{t \to +\alpha} x_i(t) = \lim_{t \to +\infty} (k_1 + k_2 \cdot \lambda_2(t))$$
$$= \frac{s(U(t) - U(t-1)) + \varphi_1 \, pbest_i(t) + \varphi_2 g \, best(t)}{\varphi} \tag{15}$$
$$\triangleq O_i$$

As the iteration progresses, $U(t)$ tends to $U(t-1)$, further deduce:

$$
\begin{aligned}
O_i &= \frac{s \cdot (U(t) - U(t-1))}{c_1 + c_2} + \frac{c_1 \cdot pbest_i(t) + c_2 \cdot gbest(t)}{c_1 + c_2} \\
&= \frac{s \cdot (U(t) - U(t-1))}{c_1 + c_2} + O_i \\
&\triangleq O_i
\end{aligned}
\tag{16}
$$

Based on the above deduction results, the improved particle swarm system can converge to a certain point in the stable area, and when it evolves to a certain generation, the stable point is close to the traditional particle swarm. It is worth noting that this conclusion is independent of the parameter s.

Improved Reverse Particle Swarm Optimization Algorithm. The traditional PSO velocity update formula is replaced with a non-inertial velocity formula, and the reverse particle swarm algorithm is improved by combining the general reverse learning (GOBL) strategy. A non-inertial reverse particle swarm optimization algorithm is designed, abbreviated as NRPSO. The main idea is as follows:

First, initialize the required parameters. This includes the initial layout of particles, the size of the particle swarm, the position and flight speed of particles in space, the number of iterations, etc. In the network element layout optimization, each particle represents a network element layout, which is composed of multiple network element objects. If the number of network elements in the layout is D, and each network element has an ID, the network element layout can be represented as $Ns = [Net_1, Net_2, \ldots, Net_D]$. If the population size is N, then the population can be represented as $P = [Ns_1, Ns_2, \ldots, Ns_N]$.

See pseudocode of NRPSO algorithm in Algorithm 1 and Algorithm 2, where N represents the swarm size, P and OP represent a particle swarm and its reverse particle swarm, D represents the dimensionality of the target, and gr represents the probability of using the GOBL strategy.

3.3 Network Element Layout Optimization Algorithm Based on Adaptive Simulated Annealing and Reverse Particle Swarm Fusion

Due to the fact that the first term in the standard PSO algorithm's particle swarm velocity update formula is an inertia term, which can maintain the diversity of the population, it also reduces the convergence speed of the algorithm. At the same time, the selection of ω has a great impact on the algorithm. Therefore, this inertia term is improved, and the non-inertia particle swarm update formula is introduced to further enhance the population's local development ability and improve the convergence speed. But, it still cannot avoid the algorithm from falling into local optimal solutions. In order to further solve this problem, the adaptive elite mutation (AEM) strategy is used to improve the global search ability of the algorithm and avoid falling into local optima.

Algorithm 1. Non-inertial Reverse Particle Swarm Optimization algorithm

---

Input: Randomly initialize a particle swarm P containing N particles, set the maximum number G of iterations, initial value T_g of iteration count, and probability gr of using the GOBL strategy. $P = [Ns_1, Ns_2, \ldots, Ns_N]$ is the population, where each vector represents a network element layout.

Output: Ns

1: **while** $T_g < G$ **do**
2:     **if** $rand < gr$ **then**
3:         RPSO()
4:     **else**
5:         **for** $i = 1 to N$ **do**
6:             According to the formula $v_i(t + 1) = s \cdot (U(t) - U(t - 1)) + c_1 rand\,(pbest_i(t) - x_i(t)) + c_2\,rand\,(gbest(t) - x_i(t))$ update velocity v_i and update velocity$x_i(t)$position according to the formula $x_i(t + 1) = x_i(t) + v_i(t + 1)$ calculate fitness for each particle according to the formula $Ef(s) = \frac{1}{n}\sum_{i=1}^{n} Error_i$ update $pbest$ and $gbest$
7:         **end for**
8:     **end if**
9:     $T_g = T_g + 1$
10: **end while**
11: **return** N_s

Algorithm 2. Reverse Particle Swarm Optimization algorithm

Input: Particle swarm P containing N particles. $P = [Ns_1, Ns_2, \ldots, Ns_N]$ is the population, where each vector represents a network element layout.

Output: Next Generation Population P Combining Reverse Particle Swarm Optimization

1: Update da,db according to $da = \min{(x_{ij})}, db = \max{(x_{ij})}$.
2: Calculate the inverse particle swarm OP of particle swarm P according to $ox_i = k(da + db) - x_i$.
3: **if** $ox_{ij} < da_j$ or $ox_{ij} > db_j$ **then**
4:     $ox_i = rand\,(da_j, db_j)$
5: **end if**
6: Calculate the fitness of each particle in the particle swarm $\{P \cup OP\}$ according to $Ef(s) = \frac{1}{n}\sum_{i=1}^{n} Error_i$
7: Select N optimal particles to form a new population P
8: Update $pbest$ and $gbest$
9: **return** P

Adaptive Elite Mutation Strategy. In order to reduce the possibility of particles falling into local optimal solutions, mutation strategies can be used to help individuals jump out of local optimal solutions, as the direction of particle flight follows the current global optimal value *gbest* of the group. AEM is used to prevent premature convergence and considers the current global best position *gbest* as the elite particle. The test position *gbest** is generated by (17):

$$gbest^* = gbest + \text{sign} \cdot \left(C + \frac{\arctan(xm)}{\pi} \right) \tag{17}$$

Among them, sign represents a symbolic constant, C indicating a constant whose value is given by (18):

$$C = \begin{cases} 1.5 & \text{st}_\text{d} < 10^{-2} \\ 1.0 & 10^{-2} \le \text{st}_\text{d} < 10^{-1} \\ 0.5 & \text{other} \end{cases} \tag{18}$$

st_d is the standard deviation of fitness, which is determined by (19).

$$st\_d = \sum_{i=1}^{N} \left| \frac{f_i - f_\text{gbest}}{f_\text{gbest}} \right| \tag{19}$$

Among them, f_gbest represents the fitness of *gbest*, f_i represents the fitness of the ith particle. $xm(i)$ represents the degree of mutation, which is determined by (20).

$$xm(i) = \left(1 - \frac{rd_i}{rd_\text{max}} \right) \cdot \exp \left(-\lambda \cdot \frac{iter}{iter_\text{max}} \right) \tag{20}$$

where λ is an uncertain constant. $iter$ and $iter_\text{max}$ represent the current iteration number and the maximum iteration number, rd_i is the distance of *gbest* and the average value of N particles' *pbest* in the i-dimension. rd_max is the maximum of rd_i.

Adaptive Simulated Annealing Non-Inertial Reverse Particle Swarm Optimization Algorithm. The non-inertial velocity formula is used instead of the traditional PSO velocity update formula, combined with the general opposition-based learning (GOBL) strategy and the adaptive elite mutation (AEM) strategy, to further improve the non-inertial reverse particle swarm optimization (NRPSO) algorithm. In the NRPSO algorithm, a fixed reverse strategy is used. Experimental results show that the performance is best when the *gr* is 0.3. However, this fixed reverse strategy is not adjusted according to the particle's flight process, which is not the best choice. Inspired by the idea of Simulated Annealing (SA), the ASA-NRPSO algorithm is designed to use the Simulated Annealing algorithm to adaptively select the reverse strategy according to the particle's needs at different stages. That is

$$gr = \exp \left(\frac{\lg(|f(U(t)) - f(U(t-1))|)}{t} \right) \tag{21}$$

$U(t)$ and $U(t-1)$ are the mean positions of all particles for the t generation and the $t-1$ generation. $f(U(t))$ and $f(U(t-1))$ are the fitness values for the corresponding mean positions. As can be seen from Eq. (21), the smaller the $gr = \exp\left(\frac{\lg(|f(U(t))-f(U(t-1))|)}{t}\right)$, the faster the convergence speed of the algorithm.

The pseudocode of the ASA-NRPSO algorithm is shown in Algorithm 3 and algorithm 4. Here, N represents the number of particles in the swarm, P and OP represent the particle swarm and its reverse particle swarm, respectively, and gr represents the probability of using the GOBL strategy.

Algorithm 3. Reverse Particle Swarm Optimization algorithm

Input: Randomly initialize a particle swarm P containing N particles, set the maximum number G of iterations, initial value T_g of iteration count, and probability gr of using the GOBL strategy.

Output: N_s

1: Calculate the mean position of each particle in the 0th generation $U(0)$ and $f(U(0))$.
2: **while** $T_g < G$ **do**
3:     **if** $rand < gr$ **then**
4:         OPSO()
5:     **else**
6:         NIV()
7:     **end if**
8:     **for** $i = 1 to N$ **do**
9:         $xm(i) = \left(1 - \frac{rd_i}{rd_{\max}}\right) \cdot \exp\left(-\lambda \cdot \frac{iter}{iter_{max}}\right)$
10:         $tg^*_{best} = gbest + sign \cdot \left(C + \frac{\arctan(xm)}{\pi}\right)$
11:     **end for**
12:     **if** $f(gbest^*)$ is better than $f(gbest)$ **then**
13:         $gbest = gbest^*$
14:     **end if**
15:     Find the mean position $U(T_g)$ and $f(U(T_g))$ of each particle in generation T_g
16:     Calculate gr according to $gr = \exp\left(\frac{\lg(|f(U(T_g))-f(U(T_g-1))|)}{t}\right)$
17:     $T_g = T_g + 1$
18: **end while**
19: **return** N_s

Algorithm 4. Non-inertia updating of particle swarm(NIU)

Input: the mean initial position $U(0)$ and $f(U(0))$ of each particle

Output: P

1: **for do**
2:     Update v_i according to $v_i(t + 1) = s \cdot (U(t) - U(t - 1)) + c_1 \text{ rand (pbest}_i(t) - x_i(t)) + c_2 \text{ rand }(gbest(t) - x_i(t))$.
3:     Update x_i according to $x_i(t + 1) = x_i(t) + v_i(t + 1)$.
4:     Calculate the fitness of each particle according to $Ef(s) = \frac{1}{n}\sum_{i=1}^{n} \text{Error }_i$
5:     Update $pbest$ and $gbest$
6: **end for**
7: **return** P

4　Experimental Results and Analysis

This section verifies the feasibility and effectiveness of the proposed network element layout optimization algorithm through experiments. The performance comparison between the network element layout fusion optimization algorithm and existing network element layout optimization algorithms is analyzed to demonstrate the effectiveness of the proposed algorithm.

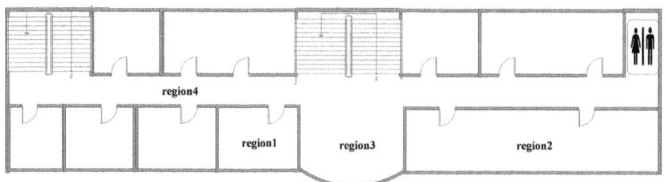

Fig. 1. Experimental scenario

4.1　Construction of Experimental Scenarios

The experimental scenarios are selected from some scenes of the School of Computer Science and Technology on the second floor of Building 21 of Harbin Engineering University.

The second floor of No. 15 Teaching Building of Qiqihar University is selected as the experimental scene, and the plane diagram is shown in Fig. 1. Area 1 is the office, Area 2 is the laboratory, Area 3 is the platform, and Area 4 is the corridor.

As region 4 are rectangular areas, signal propagation between the terminal and the network element is line-of-sight propagation, and the measured noise is relatively small. Moreover, the range of region 4 is relatively large, and the calculation and optimization costs are high, with unclear optimization effects. Region 1 is relatively small and the optimization effect is not significant. In order to better study the impact of network element layout on positioning accuracy in complex indoor environments, region 2 and region 3 are selected for performance comparison. The schematic floor plan is shown in Fig. 1.

The total length of the positioning area is 25.48 m, and the width is 12.16 m. The distance from the floor to the ceiling is 3.36 m, and the wall thickness is 0.15 m. region 3 is a rectangular area with a single curved surface, measuring 12.16 m in length and 7.6 m in width. Region 4 is a laboratory, with a length of 17.73 m and a width of 11.66 m. The main furnishings in the room include computer desks and backrest chairs. The height of the desk is 0.7 m, and the material is solid wood. There is an aluminum alloy backplane and host bracket under the desktop, which is 0.5 m high. According to the basic specifications of network element deployment, the network elements are divided into 1 m×1 m grids at a height of 2.86 m from the ground. Deploying network elements on the ceiling will produce a large amount of multipath signals, which seriously affects the positioning accuracy. Therefore, the network elements are deployed at a position 0.5 m away from the ceiling and close to the wall. There are 98 positions available for network element deployment. Since the position of the user's handheld terminal is roughly 1.2 m from the ground, several small cubes with a side length of 1 m are divided at a position 1.2 m from the ground, and the center position of the cube is used as the terminal position. There are 264 in total.

4.2   Analysis of Experimental Results

In simulation experiments, the algorithm was iterated 100 times, and the specific parameter settings are shown in Table 5.

Table 5. Terminal properties description

parameter	C_1	C_2	s	gr	N	D
value	1.496	1.496	0.2	0.3	50	30

We optimized 98 network elements and detected the positioning area coverage and average positioning error of 264 terminal positions. In the process of network element optimization configuration, the goal is to converge the average positioning error of each location to a minimum or second minimum. In the process of network element generation, optimization must be constantly carried out. Each particle is a network element layout. If fingerprint positioning is adopted in the generation process of network element layout, a fingerprint library must be established for each generation of particles. Since each iteration produces a new population, it is impossible to rebuild the fingerprint library each time. Therefore, in the process of generating network element layout in this experiment, the TDOA/AOA hybrid positioning method with better positioning effect is selected to calculate the positioning error. After the network element layout is generated, the fingerprint positioning method is used to compare the positioning performance to verify the feasibility of improving the positioning accuracy through network element layout optimization.

Experiments on Influence of Number of Deployed Network Elements on Positioning Accuracy. In this part, we selected 6–9 network elements and generated optimized network layouts to compare the impact of the number of network elements on positioning accuracy. The transmission power of the selected network elements is 23 dbm, and the frequency is 1.8 GHZ. The experimental results are shown in Fig. 2.

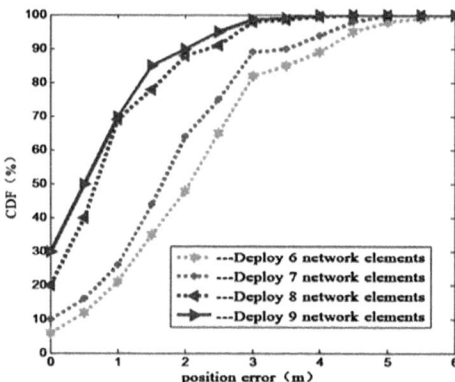

Fig. 2. The influence of the number of deployed network elements on positioning performance.

Deploying 6 network elements can result in an average positioning error of 2.82 m, 7 network elements can result in 2.18 m, 8 network elements can result in 0.968 m, and 9 network elements can result in 0.856 m. As shown in Fig. 2, the positioning error of 8 network elements is mostly within 2 m, with 69% within 1 m and 18% between 1 m and 2 m, and the increase in accuracy when deploying 9 network elements is not significant. Therefore, this experiment chose to deploy 8 network elements in the area to be located.

Experiments of First Path Signal Coverage. The main factors affecting positioning accuracy are signal coverage and signal quality. The first path signal is the signal directly received by the terminal after being transmitted by the network element, without multipath, and has the best signal quality. If the first path coverage of each reference point can be ensured, it will greatly improve the positioning quality. In this experiment, after optimizing the network element layout, the number of first path signal coverage at the terminal location is mainly concentrated between 5–7. Compared with the positioning signal coverage rate of before network element layout optimization and the ASA-GA algorithm proposed in literature [25], the first path coverage statistics are shown in Fig. 3.

As shown in Fig. 3, according to the universal network element coverage standard in K-coverage, before optimizing the network element layout, the coverage rate of the first path signal is 89.77%, while the first path signal coverage rate of other optimized algorithms reaches 100%. Literature research shows that using location fingerprinting positioning method, each positioning point can achieve ideal positioning effect by receiving signals from 5 network elements. When, before network element optimization, the first path signal coverage rate was less than 50%, and the ASA-GA algorithm could reach 95%, while the NRPSO algorithm and the ASA-NRPSO algorithm could both reach 99.

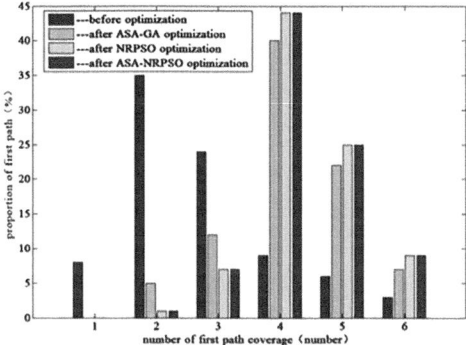

Fig. 3. The influence of the number of deployed network elements on positioning performance.

Experiments on Location Accuracy. Before optimizing the network element layout, there were about 10% of first path signal coverage gaps, which would cause a decrease in location accuracy. After optimizing the network element layout, the coverage rate of the first path signal reaches 100%, thus effectively improving the location accuracy. The experimental result also confirms this, as the coverage rate increases, the location accuracy significantly improves. The details of coverage rate and location accuracy refer to Table 6.

Table 6. Terminal properties description

	before optimization	ASA-GA	ASA-NRPSO
positioning signal coverage rate	89.77%	100%	100%
average positioning error	2.874 m	1.085 m	0.968 m

After applying the ASA-NRPSO algorithm proposed in this paper for network element layout optimization, the coverage of the first path signal was increased to 100%, which is 10.23% higher than before the optimization. The average positioning error was reduced by about 1.9 m. The cumulative distribution of positioning error before and after network element layout optimization is shown in Fig. 4. The three-dimensional positioning error distribution of region 3 after optimization by the ASA-NRPSO algorithm is shown in Fig. 5. From Figs. 4 and 5, it can be seen that using NRPSO and ASA-NRPSO optimization, about 70% of cases can obtain positioning error less than 1 m, with a probability of positioning error within 2 m at around 87%, and the maximum positioning error not exceeding 3.6 m. Compared with the network element layout optimization before, the positioning accuracy has been significantly improved.

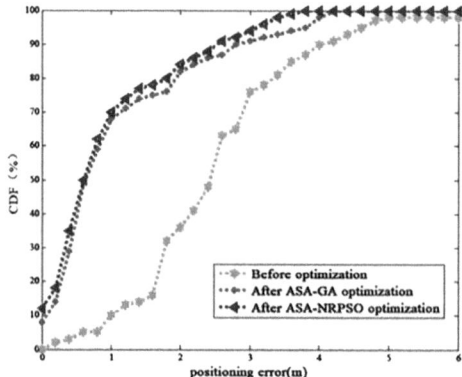

Fig. 4. The influence of the number of deployed network elements on positioning performance.

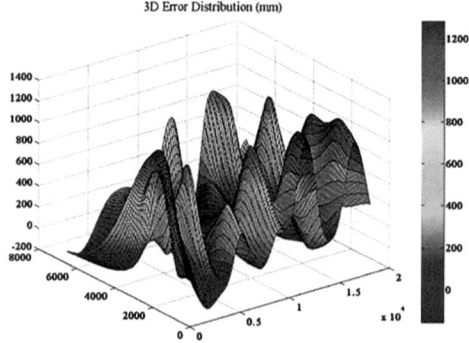

Fig. 5. The influence of the number of deployed network elements on positioning performance.

Experiments of Algorithm Convergence Speed. The positioning accuracy of NRPSO and ASA-NRPSO is basically the same. The difference between the two algorithms is in the convergence speed, as shown in Fig. 6.

From Fig. 6, it can be seen that the ASA-GA algorithm proposed in reference [25] takes about 25 iterations to achieve full coverage, NRPSO algorithm takes about 20 iterations to achieve full coverage, and ASA-NRPSO algorithm takes about 15 iterations to achieve full coverage. The convergence speed of the algorithm proposed in this paper has been significantly improved.

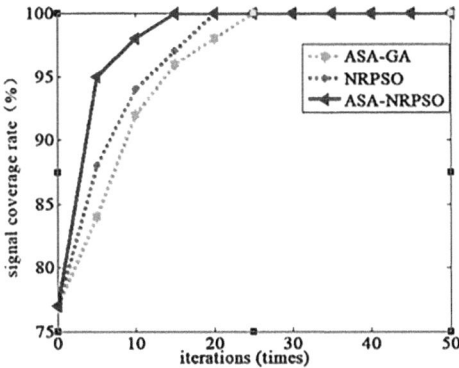

Fig. 6. The influence of the number of deployed network elements on positioning performance.

5 Conclusion

This paper proposes an indoor network element layout optimization model to address the problems of low positioning accuracy coursed by signal coverage gaps and poor positioning reference signal quality in indoor spaces. Based on this model, an ASA-NRPSO algorithm is proposed to optimize the network element layout. First, the simulated annealing idea is integrated into the traditional particle swarm algorithm to adaptively select reverse strategies based on the different needs of particles at different stages, optimize the particle swarm, and avoid particles getting stuck in local optima. Second, an non-inertia particle swarm update formula is introduced to more fully utilize group information to guide the direction of motion of the next generation of particles, further enhancing the local development ability of the population and improving the convergence speed of the particle swarm. Third, an adaptive elite mutation strategy is used to improve the algorithm's global search ability. The experiments show that after optimization by the ASA-NRPSO algorithm, the first-path signal coverage rate of the space to be located is increased to 100%, which is 10.23% higher than before network element optimization, and the average positioning error is reduced by about 1.9 m.

Acknowledgments. This research was supported by the Basic Business Project in Education Department of Heilongjiang Province of China (No. 145109140).

References

1. Agiwal, M., Roy, A., Saxena, N.: Next generation 5g wireless networks: a comprehensive survey. IEEE Commun. Surv. Tutorials **18**, 1617–1655 (2016)
2. Bais, A., Kiwan, H., Morgan, Y.L.: On optimal placement of short range base stations for indoor position estimation. J. Appl. Res. Technol. **12**, 886–897 (2014)

3. Bloodworth, C.H., et al.: Impact of simulated mitraclip on forward flow obstruction in the setting of mitral leaflet tethering: an in vitro investigation. Catheter. Cardiovasc. Interv. **92**, 1–11 (2017)

4. Bonnet, É., Escoffier, B., Paschos, V.T., Stamoulis, G.I.: Purely combinatorial approximation algorithms for maximum k-vertex cover in bipartite graphs. Discret. Optim. **27**, 26–56 (2017)

5. Chen, G., Cheng, L., Shao, R., Wang, Q., Wang, S.: A review of device-free indoor positioning for home-based care of the aged: Techniques and technologies. Computer Modeling in Engineering & Sciences (2023)

6. Chen, S., Wang, H., Chen, D., Liu, X., Hu, H.: Research on indoor network element layout optimization method for high-precision positioning. Comput. Eng. Sci. **40**, 341–347 (2018)

7. Chen, Y., Francisco, J.A., Trappe, W., Martin, R.P.: A practical approach to landmark deployment for indoor localization. In: 2006 3rd Annual IEEE Communications Society on Sensor and Ad Hoc Communications and Networks **1**, 365–373 (2006)

8. Gharghan, S.K., Nordin, R., Ismail, M., Ali, J.A.: Accurate wireless sensor localization technique based on hybrid pso-ann algorithm for indoor and outdoor track cycling. IEEE Sens. J. **16**, 529–541 (2016)

9. Han, J., Koenig, S.: A multiple surrounding point set approach using theta* algorithm on eight-neighbor grid graphs. Inf. Sci. **582**, 618–632 (2021)

10. He, H.T., Quan, Q.Y., Wang, W.B.: The necessary observation space for the lowest gdop in 2-d wireless location systems. In: 3rd International Conference on Wireless Communication and Sensor Networks (WCSN 2016), pp. 453–457. Atlantis Press (2016)

11. Islam, G.Z., Kashem, M.A.: Efficient resource allocation in the ieee 802.11ax network leveraging ofdma technology. J. King Saud Univ. Comput. Inf. Sci. **34**, 2488–2496 (2020)

12. Kalantari, E., Yanikomeroglu, H., Yongaçoğlu, A.: On the number and 3d placement of drone base stations in wireless cellular networks. In: 2016 IEEE 84th Vehicular Technology Conference (VTC-Fall), pp. 1–6 (2016)

13. Kang, L., Chen, R.S., Cao, W., Chen, Y.C.: Non-inertial opposition-based particle swarm optimization and its theoretical analysis for deep learning applications. Appl. Soft Comput. **88**, 106038 (2020)

14. Kim, D., Kim, H., Li, D., Kwon, S.S., Tokuta, A.O., Cobb, J.A.: Maximum lifetime dependable barrier-coverage in wireless sensor networks. Ad Hoc Netw. **36**, 296–307 (2016)

15. Kim, S.H., Lee, K., Kim, Y., Shin, J., Shin, S., Chong, S.: Dynamic control for on-demand interference-managed wlan infrastructures. IEEE/ACM Trans. Networking **28**, 84–97 (2020)

16. Kolodziej, K.W., Hjelm, J.: Local positioning systems: LBS applications and services. CRC Press (2017)

17. Maneerat, K., Kaemarungsi, K.: Robust system design using bilp for wireless indoor positioning systems. Mob. Inf. Syst. **2018**, 4198504:1–4198504:19 (2018)

18. Shaikh, J.A., Solano-González, J., Stojmenovic, I., Wu, J.: New metrics for dominating set based energy efficient activity scheduling in ad hoc networks. In: 28th Annual IEEE International Conference on Local Computer Networks, 2003. LCN '03. Proceedings, pp. 726–735 (2003)

19. Song, J., Jeong, H., Hur, S., Park, Y.: Improved indoor position estimation algorithm based on geo-magnetism intensity. In: 2014 International Conference on Indoor Positioning and Indoor Navigation (IPIN), pp. 741–744 (2014)

20. Valenzuela-Valdés, J.F., Padilla, J.L., Padilla, P., Luna, F., Fernández-González, J.M.: Design rules for antenna placement on mimo system. J. Electromagnetic Waves Appl. **30**, 1731–1739 (2016)
21. Wang, H.Q., Liu, X.B., Lü, H.W., Feng, G.S., Yang, Y.P.: Method of diamond supplement for indoor location micro base station placement. Journal of Beijing University of Posts and Telecommunications (2018)
22. Wang, X., Wang, Y., Dang, Z., Pei, H., Zhang, L.: An improved toa model based on error correction and self-genetic algorithm. Int. J. Performability Eng. **14**, 2374 (2018)
23. Yan, L., Mao, Y.: Wireless location technology of gauss particle filter under nlos environment. In: Proceedings of the 2016 3rd International Conference on Materials Engineering, Manufacturing Technology and Control, pp. 251–256. Atlantis Press (2016)
24. Yu, C.W.: Randomized coverage algorithms for mobile sensor networks. In: Workshop on Combinatorial Mathematics and Computation Theory. Citeseer (2008)
25. min Yu, X., qiang Wang, H., Lv, H., Liu, X., qiu Wu, J.: A fusion optimization algorithm of network element layout for indoor positioning. EURASIP J. Wirel. Commun. Networking **2019**, 1–12 (2019)
26. Zhang, Z., Zhou, J., chang Mo, Y., Du, D.Z.: Performance-guaranteed approximation algorithm for fault-tolerant connected dominating set in wireless networks. In: IEEE INFOCOM 2016 - The 35th Annual IEEE International Conference on Computer Communications, pp. 1–8 (2016)
27. Zhou, B., Tu, W., Mai, K., Xue, W., Ma, W., Li, Q.: A novel access point placement method for wifi fingerprinting considering existing aps. IEEE Wireless Commun. Lett. **9**, 1799–1802 (2020)
28. Zhou, J., Shi, J., Qu, X.: Landmark placement for wireless localization in rectangular-shaped industrial facilities. IEEE Trans. Veh. Technol. **59**, 3081–3090 (2010)

An Efficient Lattice-Based Authentication Protocol for the Vehicular Ad Hoc Network

Xinyong Chen[1], Jiageng Chen[1(✉)], Jinquan Luo[2], and Hongwei Liu[2]

[1] School of Computer Science, and Hubei Provincial Key Laboratory of Artificial Intelligence and Smart Learning, Central China Normal University, Wuhan 430079, Hubei, China
chinkako@gmail.com
[2] School of Mathematics and Statistics, Central China Normal University, Wuhan 430079, Hubei, China

Abstract. Vehicular Ad Hoc Networks (VANETs) are crucial for intelligent transportation systems (ITS), yet they face significant security and privacy challenges. The open wireless environment makes VANETs vulnerable to attacks, and existing authentication protocols often have limitations, such as susceptibility to quantum threats or high communication overhead. In this paper, we propose an effective privacy-preserving authentication protocol using lattice-based group signatures for VANETs. Our protocol ensures quantum security and provides robust security features. It not only ensures privacy protection for vehicles but also enables efficient tracing of malicious signers when necessary. In addition, we show that the proposed protocol features small and fixed public key and signature sizes, independent of the number of members, and incurs low communication overhead.

Keywords: Privacy-preserving · Lattice · Vehicular ad hoc networks · Group signature · Authentication

1 Introduction

Recently, with the advancement of IoT and wireless sensor technology, VANETs have garnered extensive attention as a crucial component of ITS. In ITS scenarios, vehicles act as information disseminators, broadcasting traffic information via VANETs to support driving strategy adjustments, route planning, and the formation of a complex traffic information network. This network is essential for the efficient operation of ITS. Within this standard, the entities involved in the VANET system fall into two distinct classifications: On-Board Units (OBUs) and Road-Side Units (RSUs). Ordinarily, any vehicle is outfitted with OBU for broadcasting messages and processing received messages. RSUs, serving as stationary base stations along the roadway, furnish vehicles with internet connectivity and traffic-related data. The OBU embedded in a vehicle can interact with other

© The Author(s), under exclusive license to Springer Nature Switzerland AG 2025
W. Meng et al. (Eds.): ADIoT 2024, LNCS 15397, pp. 76–89, 2025.
https://doi.org/10.1007/978-3-031-85593-1_5

vehicles (Vehicles-to-Vehicles, V2V) and RSUs (Vehicles-to-Infrastructures, V2I) [1,14,16]. Therefore, in VANETs, drivers can obtain traffic information through the reception of broadcast messages transmitted by other vehicles or RSUs. The communication in V2V or V2I can be realized by the Dedicated Short Range Communication (DSRC) protocol [7]. Furthermore, in practice, it is usually necessary to introduce a Trusted Authority (TA) to supervise and coordinate the entire VANETs. An example of VANETs architecture is displayed in Fig. 1.

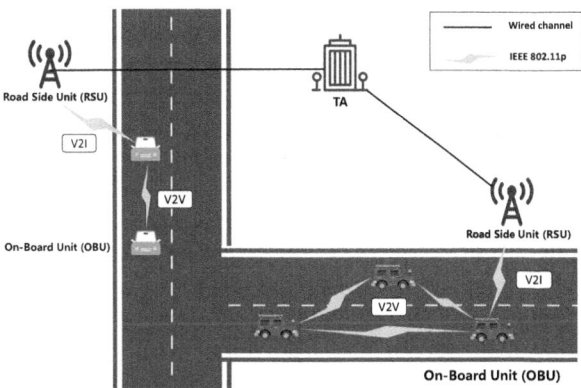

Fig. 1. An instance of VANETs.

VANETs' development in modern transportation brings convenience but also security issues. The open environment allows attackers to forge or tamper messages, which may mislead driving decisions and threaten safety. Hence, message authentication is needed, along with user privacy protection. However, this may enable malicious vehicles. Thus, TA can track the real identity of the signer when necessary. During the process of seeking solutions to meet the above-mentioned security and privacy requirements, various cryptographic techniques and authentication protocols have emerged. Among them, group signatures have shown great potential in the field of anonymous authentication of the Internet of Vehicles with their unique properties.

Group signatures enable members to sign messages in the name of the entire group without revealing the signer's information. In VANETs, vehicles, as group members, enjoy anonymity when signing messages. The group administrator, acting as a TA, ensures the system remains traceable. The TA can track the identity of a signer if malicious behavior is detected, balancing privacy and security needs. Therefore, group signatures provide a valuable foundation for developing secure and anonymous authentication protocols in VANETs.

Furthermore, quantum computing's progress endangers traditional public key cryptosystems. Most VANETs auth protocols rely on classical hard problems, but Shor's algorithm on quantum computers can solve them. Hence, a quantum-resistant VANETs auth scheme is essential.

Our Contribution. This paper puts forward an effective privacy-protecting authentication protocol designed for VANETs. Our protocol is based on **M-SIS** problem and **M-LWE** problem, both of which have been theoretically proven to be resistant to quantum computing attacks. By employing these advanced cryptographic techniques, our protocol not only ensures the security of communication but also provides a strong guarantee for user privacy.

Our scheme possesses the following characteristics: correctness, anonymity, traceability, and resistance to replay attacks. These features ensure the reliability and security of the system while protecting user privacy. Correctness guarantees that all legitimate message signatures can be accurately verified. Anonymity ensures that the identity of participants remains confidential. Traceability allows message signatures to be traced when necessary for auditing or investigation. Resistance to replay attacks prevents old messages from being maliciously reused, thus maintaining the overall security of the system.

Additionally, we present the parameter settings of our protocol to achieve 128-bit security. In the protocol we designed, a single group can accommodate up to 2^{23} members. Our protocol's group public key size and the signature size are 127.0 KB and 93.5 KB respectively, and they will not increase with the growth of the number of members. Thus, our protocol incurs low communication overhead, ensuring high efficiency. Such a scale setting endows it with stronger adaptability and suitability in the highly vehicle-dense VANETs scenarios.

1.1   Related Works

To tackle privacy and security challenges in VANETs, Hubaux et al. [5] introduced the key idea of anonymous authentication in VANETs. Subsequently, a large number of protocols emerged as the times require. These protocols mainly address security threats and support privacy-protection authentication in VANETs. In this environment, leveraging special digital signature primitives like ring and group signatures is an effective means to achieve privacy-protected message integrity authentication.

Under the ring signature mechanism, any user can perform the signing operation for any message in the name of the whole ring. Furthermore, any verifier who acquires the ring public key has the ability to verify whether the signature is from this ring. The reference [6,13] proposed authentication schemes for VANETs relying on ring signatures, achieving quantum security and unconditional anonymity. However, TA in these schemes cannot trace vehicles which carry out malicious behaviors. In practice, the TA should be able to know the real identity of vehicles when the situation calls for it [4,19]. To address this problem, based on the ring signature, the scheme [12] introduced a pseudonyms mechanism, which enables TA to track the real identity associated with it according to the pseudonyms.

Group signature, allowing group members to sign messages on behalf of the entire group, is also an important cryptographic primitive for privacy protection. In group signatures, there is a group administrator role (corresponding to the TA), who is responsible for managing group members and can trace the identity of the signer. The references [9,20,22] presented VANETs authentication protocols based on group signatures, achieving message anonymity and reserving the right of the TA to trace malicious users. In addition, in the reference [15], scholars introduced a threshold anonymous authentication protocol using group signatures and use batch message processing technology to accelerate message verification between vehicles. Recently, Cao et al. [3] presented an authentication protocol using lattices, which achieved forward security.

Organization. We introduce preliminaries in Sect. 2. The system model and security requirements are described in Sect. 3 and the proposed scheme is described in Sect. 4. Next, Sect. 5 introduces the supporting ZKAoK. We analyze the correctness and security of our scheme in Sect. 6. The evaluation and comparison details are in the Sect. 7. Finally, we conclude our work in Sect. 8.

2   Preliminaries

In this section, we have elaborated and defined some basic knowledge which will be used in this paper.

2.1   Lattice and Hard Problems

Our protocol relies on lattice-based cryptography, and here we first present the basic algebraic structure of lattice. For positive integers q and n, $\mathbf{Z} \in \mathbb{Z}_q^{n \times m}$ and $\mathbf{t} \in \mathbb{Z}_q^n$, define the full-rank m-dimensional integer lattices as follow:

$$\Lambda_q(\mathbf{Z}) = \{\mathbf{v} \in \mathbb{Z}_q^m \mid \exists \mathbf{y} \in \mathbb{Z}_q^n \ \ s.t. \ \ \mathbf{v} = \mathbf{Z}^{\mathrm{T}} \cdot \mathbf{y} \bmod q\}$$
$$\Lambda_q^{\perp}(\mathbf{Z}) = \{\mathbf{v} \in \mathbb{Z}_q^m \mid \mathbf{Z} \cdot \mathbf{v} = \mathbf{0} \bmod q\}$$
$$\Lambda_q^{\mathbf{t}}(\mathbf{Z}) = \{\mathbf{v} \in \mathbb{Z}_q^m \mid \mathbf{Z} \cdot \mathbf{v} = \mathbf{t} \bmod q\}$$

Let q be a prime and $\mathbb{Z}_q = \left[-\frac{q-1}{2}, \frac{q-1}{2}\right]$. Let d be a power of 2, the ring $R = \mathbb{Z}[X]/(X^d + 1)$, $R_q = \mathbb{Z}_q[X]/(X^d + 1)$ and $R_{kd,q} = \mathbb{Z}_q[X]/(X^{kd} + 1)$.

$\|\cdot\|$ represents the Euclidean norm, $\|\cdot\|_\infty$ represents the infinity norm, and $s_1(\cdot)$ represents the spectral norm. For $r = r_0 + r_1 \cdot X + ... + r_{d-1} \cdot X^{d-1} \in R$, we can acquire that $\|r\| = \sqrt{k_0^2 + r_1^2 + ... + r_{d-1}^2}$, $\|r\|_\infty = \max_i(|r_i|)$. For a vector of ring elements $\mathbf{k} = (k_1, ..., k_m)^{\mathrm{T}}$, we set $\|\mathbf{k}\|_\infty = \max_j(\|k_j\|_\infty)$.

Definition 1 (M-SIS [8]). *For a uniformly random $\mathbf{A} \in R_q^{n \times m}$, the* ***M-SIS****$_{q,n,m,\beta}$ problem (over an implicit ring R) is to find a vector $\mathbf{u} = (u_1, ..., u_m)^{\mathrm{T}} \in R^m$, satisfying $0 < \|\mathbf{u}\| \le \beta$ and $\mathbf{A} \cdot \mathbf{u} = 0$.*

Definition 2 (M-LWE [2,8]). *Given a error distribution χ over R and $\mathbf{z} \in \chi^n$, define a distribution $T_{\mathbf{z},\chi}$ outputs the pair $(\mathbf{B}, \mathbf{B} \cdot \mathbf{z} + \mathbf{e}) \in R_q^{m \times n} \times R_q^m$, where $\mathbf{B} \leftarrow R_q^{m \times n}$ and $\mathbf{e} \leftarrow \chi^m$. The* ***M-LWE****$_{q,m,n}$ problem requires to distinguish m samples selected from $T_{\mathbf{z},\chi}$ and m samples selected from the uniform distribution in $R_q^{m \times n} \times R_q^m$.*

2.2 Probability Distributions and Sampling Algorithms

Definition 3. For $\sigma > 0$ and a vector $\mathbf{c} \in \mathbb{R}^n$, let $\rho_{\mathbf{c},\sigma}(\mathbf{z}) = \exp(-\pi||\mathbf{z} - \mathbf{c}||/\sigma^2)$ be a Gaussian function. For a lattice Λ, let $\rho_{\mathbf{c},\sigma}(\Lambda) = \sum_{\mathbf{z} \in \Lambda} \rho_{\mathbf{c},\sigma}(\mathbf{z})$. For any $\mathbf{y} \in \Lambda$, $D_{\Lambda,\mathbf{c},\sigma} = \rho_{\mathbf{c},\sigma}(\mathbf{y})/\rho_{\mathbf{c},\sigma}(\Lambda)$ is defined as a the discrete Gaussian distribution over Λ. In particular, we denote the distribution centered at $\mathbf{c} = \mathbf{0}$ by $D_{\Lambda,\sigma}$.

Definition 3. Bin_z represents a binomial distribution with a positive integer parameter z. Which is defined as the distribution $\sum_{j=1}^{z}(u_j - v_j)$, where $u_j, v_j \leftarrow \{0,1\}$.

Theorem 1 [11]. *There is an algorithm* $\mathsf{SampleD}(\mathbf{F}, \mathbf{u}, \mathbf{T}, \sigma)$ *which takes a matrix* $\mathbf{F} \in \mathbb{Z}_q^{n \times (m+kn)}$, *a target image* $\mathbf{u} \in \mathbb{Z}_q^n$, *a* $\mathbf{G}$*-trapdoor* $\mathbf{T} \in \mathbb{Z}_q^{m \times kn}$ *for* $\mathbf{F}$, *and parameter* $\sigma > \omega(\sqrt{\log n}) \cdot s_1(\mathbf{T})$ *as inputs, then outputs a sample from the distribution* $\mathbb{D}_{\Lambda_q^{\mathbf{u}}(\mathbf{F}),\sigma}$.

3 System Model and Security Requirements

In this section, we will briefly introduce the system model of our scheme, the security and privacy requirements.

3.1 System Model

In our protocol, the participating parties include Trusted Authority (TA), Local Trusted Authority (LTA), Road Side Unit (RSU), and On-Board Unit (OBU).

TA: As a trusted party, TA, connected to LTA via a wired channel, has relatively high administrative privileges. TA is in charge of producing the system parameters and distributing the issuing keys to each LTA.

LTA: For convenient management, TA partitions network areas into groups. Each group has an LTA as admin. LTA, a subordinate of TA, manages its area, relieving TA's pressure. LTA uses issuing key to give vehicles signing keys. If a vehicle broadcasts illegal messages, LTA uses tracing key to uncover its identity and report to TA.

RSU: Generally, RSU is an infrastructure deployed on both sides of the road and communicates with nearby vehicles by means of radio communication (DSRC protocol). RSU can conduct two-way data interaction with nearby vehicles, that is, providing services to vehicles or receiving information broadcast by vehicles.

OBU: OBUs, embedded in vehicles, are integral components in VANETs. They enable vehicles to communicate with nearby RSUs (V2I) and other vehicles (V2V) via the DSRC protocol. This communication allows vehicles to obtain network services, share and authenticate messages, and ultimately enhance the driving experience.

In Fig. 2, we detail the communication pattern of our proposed scheme to make our message authentication protocol in VANETs more understandable.

When a vehicle enters a group, it sends its real identity in a registration request to an RSU. The RSU forwards it to the area's LTA. LTA checks the info's legality, gives the vehicle a signing key using the issuing key. The vehicle then signs and broadcasts messages. Nearby vehicles verify with the group public key but can't identify the signer. If a vehicle broadcasts an illegal message, LTA traces the signer's identity with the tracing key and reports to TA.

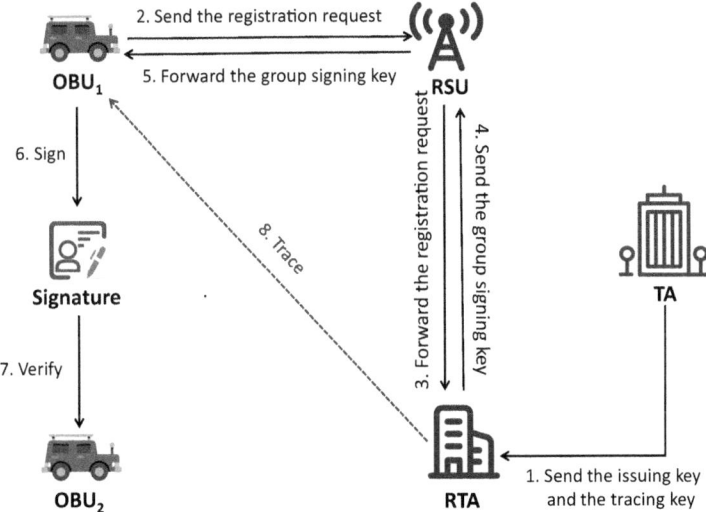

Fig. 2. The communication pattern of our scheme

3.2 Security Requirements

In VANETs, info security and privacy are crucial for wide use and development. We aim to create a secure, efficient, quantum-resistant anonymous auth protocol. It must meet specific security and privacy needs.

1. Message integrity and authentication: Before a vehicle (equipped with an OBU) or RSU takes in a message, the signature of the message must be verified to ensure it originates from other authorized members within the VANET and has not been modified during transmission.
2. Anonymity: Authorized members within the VANET can sign messages anonymously. Additionally, no party other than the TA/LTA should be able to determine the actual identity of the signer.
3. Traceability: TA/LTA can trace the actual identity of the signer by means of a valid signature.

4. Resistance to replay attack: It means that the system has robust mechanisms to detect and discard any previously transmitted messages that are resent by malicious entities, thereby ensuring the integrity and security of communication among V2V and V2I.

4   The Proposed Scheme

4.1   System Setup and Key Generation

At this stage, TA will generate system parameters and execute the key generation algorithm. The details are as follows:

– TA chooses an identity space $\mathcal{I} = \{i(X^k) \in R_{kd,p} : i \in \mathcal{B} \wedge \|i\|_1 = w\}$, where $\mathcal{B}$ is the set of non-zero binary polynomials in R_p.
– Sample $\mathbf{A} \leftarrow R_{kd,p}^{N \times (N+M)}$, $\mathbf{B}' \leftarrow R_{kd,p}^{N \times \tau N}$ amd randomness matrix $\mathbf{R} \leftarrow \{a \in R_{kd} : \|a\|_\infty \leq 1\}^{(N+M) \times \tau N}$. Then compute $\mathbf{B} = \mathbf{AR}$.
– Sample $(\mathbf{s}_{gm}^1, \mathbf{s}_{gm}^2, \mathbf{s}_{gm}^3) \leftarrow D_\sigma^{N+M} \times D_\sigma^{\tau N} \times D_\sigma^{\tau N}$. Then compute $\mathbf{u} = [\mathbf{A}|\mathbf{B}|\mathbf{B}'] \begin{bmatrix} \mathbf{s}_{gm}^1 \\ \mathbf{s}_{gm}^2 \\ \mathbf{s}_{gm}^3 \end{bmatrix}$.
– Sample $(\mathbf{A_e}, \mathbf{s_e}, \mathbf{x_e}) \leftarrow R_{p_e}^{N_e \times K_e} \times Bin_2^{N_e d} \times Bin_2^{K_e d}$. Then compute $\mathbf{b_e} = \mathbf{A_e^T s_e} + \mathbf{x_e}$.
– Select a hash function $\mathcal{H} : \{0,1\}^* \to R^{N_e \times K_e}$.
– Next, TA will send $\mathbf{R}$ and $(\mathbf{s_e}, \mathbf{x_e})$ to LTA, respectively as the group issuing key ik and the tracing key TK.
– Finally, TA broadcasts the group public key $gpk = (\mathbf{A}, \mathbf{B}, \mathbf{B}', \mathbf{u}, \mathbf{A_e}, \mathbf{b_e})$ to the whole network through RSUs.

4.2   Joining Phase

When a new vehicle $i \in \mathcal{I}$ with an embedded OBU moves into the group range, it sends a joining request to the RSU. Then the RSU forwards the request to LTA. After verifying the identity, the LTA utilizes the issuing key ik $= \mathbf{R}$ to produce a dedicated signing key for the vehicle i. The steps are as follows:

– LTA samples $\mathbf{s}_i^3 \leftarrow D_\sigma^{\tau N}$ and computes $\mathbf{y} = \mathbf{B}' \cdot \mathbf{s}_i^3$.
– Using the trapdoor sampling algorithm [11], generate short vectors $(\mathbf{s}_i^1, \mathbf{s}_i^2) \leftarrow$ SampleD$(\mathbf{A}, \mathbf{u} - \mathbf{y}, \mathbf{R}, \sigma)$ satisfying

$$\begin{cases} [\mathbf{A}|\mathbf{B} + i\mathbf{G}|\mathbf{B}'] \begin{bmatrix} \mathbf{s}_i^1 \\ \mathbf{s}_i^2 \\ \mathbf{s}_i^3 \end{bmatrix} = \mathbf{u}, \\ \left\| \begin{bmatrix} \mathbf{s}_i^1 \\ \mathbf{s}_i^2 \\ \mathbf{s}_i^3 \end{bmatrix} \right\| \leq B, \end{cases} \tag{1}$$

where $\mathbf{G}$ is a gadget matrix.

Finally, the vehicle i obtains the group signing key $gsk_i = (\mathbf{s}_i^1, \mathbf{s}_i^2, \mathbf{s}_i^3)$.

4.3   Message Signing

On input the group public key gpk, the group signature key gsk_i and the message $M \in \{0,1\}^*$, the signer executes the following steps:

- Yield the current timestamp t_s by using the system clock.
- Sample $\zeta \leftarrow R^{N_e}$, then compute $\mathbf{P} = \mathcal{H}(gpk, M, \zeta, t_s)$.
- Next, it needs to use $(\mathbf{A_e}, \mathbf{b_e})$ to encrypt its identity. Firstly, sample small elements $\mathbf{r} \leftarrow Bin_2^{K_e d}$, then compute the following ciphertext:

$$\begin{bmatrix} \mathbf{t_0} \\ \mathbf{t_1} \end{bmatrix} = \begin{bmatrix} \mathbf{A_e} \\ \mathbf{b_e^T} \end{bmatrix} \cdot \mathbf{r} + \begin{bmatrix} \mathbf{0} \\ \lfloor \frac{p_e}{2} \rfloor \cdot i \end{bmatrix}, \tag{2}$$

- Compute the following equation:

$$\tilde{\mathbf{v}} = \mathbf{Pr}. \tag{3}$$

- Subsequently, produce a non-interactive zero-knowledge argument of knowledge Π (the specific details are in Sect. 5) to manifest the possession of a secret tuple $\varphi = (\mathbf{s_i^1}, \mathbf{s_i^2}, \mathbf{s_i^3}, \mathbf{r}, i)$ satisfying the following conditions:
  1. equation (1) holds and i has binary coefficients.
  2. equation (2) holds and $\mathbf{r}$ is small.
  3. equation (3) holds.
- Finally, broadcast the message signature $\Sigma = (\zeta, \tilde{\mathbf{v}}, \mathbf{t_0}, \mathbf{t_1}, \Pi, t_s)$ to the network.

4.4   Signature Verification

On receiving the message M, the signature Σ, the verifiers (OBUs or RSUs), given the group public key gpk, perform the following steps:

- Parses the group signature $\Sigma = (\zeta, \tilde{\mathbf{v}}, \mathbf{t_0}, \mathbf{t_1}, \Pi, t_s)$.
- Compute $\mathbf{P} = \mathcal{H}(gpk, M, \zeta, t_s)$ and check the proof Π. If Π is invalid, then this algorithm return 0. Otherwise, Return 1.

4.5   Signature Tracing

To hold malicious network vehicles accountable, LTA uses the tracing key to recover the signer's identity hidden in the group signature. On input the group public key gpk, the group tracing key TK and the signature Σ, the LTA executes the following steps:

- Parses the group signature $\Sigma = (\zeta, \tilde{\mathbf{v}}, \mathbf{t_0}, \mathbf{t_1}, \Pi, t_s)$ and the group tracing key $\mathsf{TK} = (\mathbf{s_e}, \mathbf{x_e})$.
- Utilize $(\mathbf{s_e}, \mathbf{x_e})$ to decrypt $(\mathbf{t_0}, \mathbf{t_1})$ as follows.
- Compute $i' = \mathbf{t_1} - \mathbf{s_e^T} \mathbf{t_0}$. For each coefficient of i', if it is closer to 0 than to $\lfloor \frac{p_e}{2} \rfloor$, then round it to 0; otherwise, round it to 1.
- Return i'.

5   The Underlying Zero-Knowledge Argument System

Next, we elaborate on the ZKAoK used by vehicles with built-in OBUs when producing a group signature. Concretely, the prover attempts to convince the verifier of the following facts:

1. The vehicle $i \in \mathcal{I}$ is a legitimate group user who has the group signing key $gsk_i = (\mathbf{s}_i^1, \mathbf{s}_i^2, \mathbf{s}_i^3)$ issued by the LTA, satisfying Eq. (1).
2. The tuple $(\mathbf{t_0}, \mathbf{t_1})$ is the result of correctly encrypting the identity i, under the public key $(\mathbf{A_e}, \mathbf{b_e})$. The vehicle i needs to prove that Eq. (2) holds.
3. The Eq. (3) holds.

To sum up, the prover, who knows a secret witness $\varphi = (\mathbf{s}_i^1, \mathbf{s}_i^2, \mathbf{s}_i^3, \mathbf{r}, i)$, needs to prove the following relation:

$$
\begin{cases}
[\mathbf{A}|\mathbf{B} + i\mathbf{G}|\mathbf{B}'] \begin{bmatrix} \mathbf{s}_i^1 \\ \mathbf{s}_i^2 \\ \mathbf{s}_i^3 \end{bmatrix} = \mathbf{u}, \\
\tilde{\mathbf{v}} = \mathbf{Pr}, \\
\mathbf{t_0} = \mathbf{A_e r}, \\
\mathbf{t_1} = \mathbf{b}_e^T \mathbf{r} + \lfloor \frac{p_e}{2} \rfloor \cdot i, \\
\left\| \begin{bmatrix} \mathbf{s}_i^1 \\ \mathbf{s}_i^2 \\ \mathbf{s}_i^3 \end{bmatrix} \right\| \leq B, \\
\|\mathbf{r}\| \leq B_r, \\
i \in \mathcal{I}.
\end{cases}
\tag{4}
$$

For this purpose, our work makes use of the recent zero-knowledge protocol based on lattices formulated by Lyubashevsky et al. [10] (refer to it as LNP22) to prove the aforementioned relation (4), and transform it into non-interactive using the Fiat-Shamir transformation. In Fig. 3, we present the instantiation of our protocol to adapt to LNP22 protocol, maintaining the consistency of the notation with the protocol. In this instantiation, $(\mathbf{E_1}, \mathbf{v_1}, \beta_1^{(e)})$ is utilized to demonstrate exactly that $\left\| \begin{bmatrix} \mathbf{s}_i^1 \\ \mathbf{s}_i^2 \\ \mathbf{s}_i^3 \end{bmatrix} \right\| \leq B$; $(\mathbf{E_2}, \mathbf{v_2}, \beta_2^{(e)})$ is utilized to demonstrate exactly that $\|\mathbf{r}\| \leq B_r$. Besides, the tuple $(\mathbf{E_{bin}}, \mathbf{v_{bin}})$ is employed to demonstrate that the coefficients of i are all binary. In the end, the Signing algorithm of our scheme can produce a non-interactive ZKAoK Π based on this protocol.

Variable	Descriptions	Instantiation
ρ	# of equations to prove	N
ρ_{eval}	# of evaluations with const. coeff. zero	1
v_e	# of exact norm proofs	3
v_d	# non-exact norm proofs	1
k_{bin}	length of the binary vector to prove	1
$\mathbf{s_1}$	committed message in the Ajtai part	$(\mathbf{s}_i^1, \mathbf{s}_i^2, \mathbf{s}_i^3, \mathbf{r}, i)$
$\mathbf{m}$	committed message in the BDLOP part	$\emptyset$ (no message)
$f_1, \cdots f_N$	equations to prove (q is the proof system modulus)	$\frac{q}{p}[\mathbf{A}\vert\mathbf{B}+i\mathbf{G}\vert\mathbf{B}']\begin{bmatrix}\mathbf{s}_i^1\\\mathbf{s}_i^2\\\mathbf{s}_i^3\end{bmatrix}=\frac{q}{p}\mathbf{u}$
F_1	evaluation to prove const coeff. zero	$\sigma_{-1}(\sum_{i=0}^{d-1}X^i)\cdot i - \omega$
$\mathbf{E_1}$	public matrix for proving $\Vert\mathbf{E_1 s}-\mathbf{v_1}\Vert\le\beta_1^{(e)}$	$[\mathbf{I}_{N+M+2\tau N}\ \mathbf{0}\ \mathbf{0}]$
$\mathbf{v_1}$	public vector for proving $\Vert\mathbf{E_1 s}-\mathbf{v_1}\Vert\le\beta_1^{(e)}$	$\mathbf{0}$
$\beta_1^{(e)}$	upper-bound on $\Vert\mathbf{E_1 s}-\mathbf{v_1}\Vert\le\beta_1^{(e)}$	B
$\mathbf{E_2}$	public matrix for proving $\Vert\mathbf{E_2 s}-\mathbf{v_2}\Vert\le\beta_2^{(e)}$	$[\mathbf{0}\ \mathbf{0}\ \mathbf{0}\ \mathbf{I}_{K_e}\ \mathbf{0}]$
$\mathbf{v_2}$	public vector for proving $\Vert\mathbf{E_2 s}-\mathbf{v_2}\Vert\le\beta_2^{(e)}$	$\mathbf{0}$
$\beta_2^{(e)}$	upper-bound on $\Vert\mathbf{E_2 s}-\mathbf{v_2}\Vert\le\beta_2^{(e)}$	B_r
$\mathbf{D_1}$	public matrix for proving $\Vert\mathbf{D_1 s}-\mathbf{u_1}\Vert\le\beta_1^{(d)}$	$p_e^{-1}\cdot\begin{bmatrix}\mathbf{0}\ \mathbf{0}\ \mathbf{0}\ \mathbf{A_e}\ \mathbf{0}\\\mathbf{0}\ \mathbf{0}\ \mathbf{0}\ \mathbf{b}_e^{T}\ \lfloor\frac{p_e}{2}\rfloor\end{bmatrix}$
$\mathbf{u_1}$	public vector for proving $\Vert\mathbf{D_1 s}-\mathbf{u_1}\Vert\le\beta_1^{(d)}$	$p_e^{-1}\cdot\begin{bmatrix}\mathbf{t_0}\\\mathbf{t_1}\end{bmatrix}$
$\beta_1^{(d)}$	upper-bound on $\Vert\mathbf{D_1 s}-\mathbf{u_1}\Vert\le\beta_1^{(d)}$	$B_{v,enc}$
$\mathbf{E_{bin}}$	matrix for proving binary	$[\mathbf{0}\ \mathbf{0}\ \mathbf{0}\ \mathbf{0}\ 1]$
$\mathbf{v_{bin}}$	matrix for proving binary	0

Fig. 3. Instantiation of the LNP22 framework for proving Eq. (4), employing the same notation as in [10, Sect. 6].

6 Security Analysis

In this section, we will introduce the security analysis of the proposed scheme, including correctness, anonymity, traceability and resistance to replay attack.

6.1 Correctness

Our protocol's correctness depends on the facts: (1) the zero-knowledge argument of knowledge Π are perfectly complete; (2) the encryption algorithm used in the message signing phase is correct. Therefore, the signature generated by the honest vehicle i can pass the verification.

6.2 Anonymity

The anonymity requirement of our protocol stipulates that, except for the TA/LTA, cannot acquire the actual identity of the signer via the signature. When the attacker receives the signature $\Sigma = (\zeta, \tilde{\mathbf{v}}, \mathbf{t_0}, \mathbf{t_1}, \Pi, t_s)$, due to the zero-knowledge property of ZKAoK used in our protocol, he or she cannot obtain any additional information about the secret $\varphi = (\mathbf{s}_i^1, \mathbf{s}_i^2, \mathbf{s}_i^3, \mathbf{r},\ i)$ from the proof Π. In addition, the attacker are unable to obtain the encrypted content, that is, the actual identity of the signer, from the ciphertext $(\mathbf{t_0}, \mathbf{t_1})$ generated by the encryption scheme based on **M-LWE** hard problem. Thus, our protocol satisfies anonymity.

6.3   Traceability

The traceability requirement of our protocol dictates that the TA/LTA can recover the actual identity of the signer hidden in the signature when necessary.

Specifically, using TK, LTA decrypts the ciphertext $(\mathbf{t_0}, \mathbf{t_1})$ included in the signature:

$$i' = \mathbf{t_1} - \mathbf{s_e^T} \mathbf{t_0}$$
$$= \mathbf{b_e^T} \mathbf{r} + \lfloor \frac{p_e}{2} \rfloor \cdot i - \mathbf{s_e^T} \mathbf{A_e} \mathbf{r}$$
$$= \lfloor \frac{p_e}{2} \rfloor \cdot i + \mathbf{x_e^T} \mathbf{r}.$$

Under the parameter settings of our protocol, we can obtain that $\|\mathbf{x_e^T}\| < \frac{p_e}{4}$. Therefore, through the rounding operation (as shown in Subsect. 4.5), i' is restored to the actual identity i of the signer.

Table 1. Security properties comparisons

Schemes	Authentication	Anonymity	Traceability	Resistance to replay attack	Postquantum security
[21]	✓	✓	✓	✓	✗
[18]	✓	✓	✓	✗	✗
[3]	✓	✓	✓	✗	✓
[17]	✓	✓	✓	✓	✓
Ours	✓	✓	✓	✓	✓

6.4   Resistance to Replay Attack

In our VANETs protocol, any message broadcast by OBU will be accompanied by a timestamp. Once other OBUs or RSUs detect that the message has expired, then it will be discarded before the verification process. If an attacker attempts to replace the original timestamp t_s with a new one t_s', then in the signature verification algorithm, the verifier will calculate $\mathbf{P'} = \mathcal{H}(gpk, M, \zeta, t_s')$ instead of the original $\mathbf{P} = \mathcal{H}(gpk, M, \zeta, t_s)$. Since the attacker doesn't know the witness $\mathbf{r}$ to produce a legitimate proof to demonstrate that $\tilde{\mathbf{v}}' = \mathbf{P'r}$. Therefore, the message with the forged timestamp fails the verification.

7   Evaluation and Comparison

7.1   Security Comparison

In Table 1, we show the comparison of the security requirements of our scheme with those of other authentication schemes. All the schemes in the table satisfy anonymity and traceability. Our scheme not only satisfies anonymity, traceability and resistance to replay attack, but also achieves quantum-resistant security based on lattices.

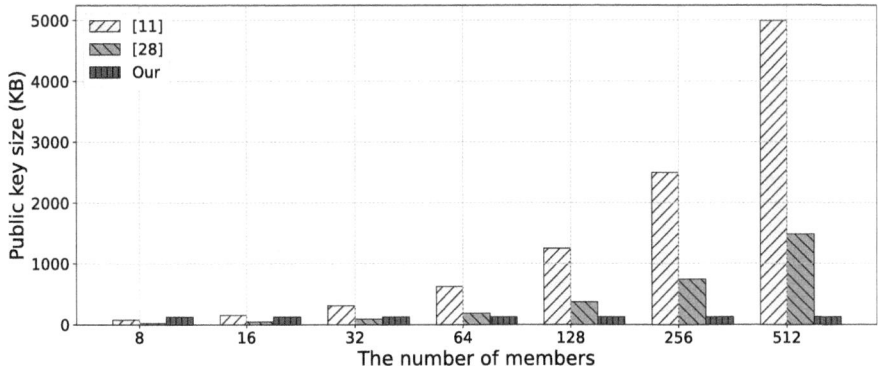

Fig. 4. Comparison of public key sizes

7.2   Parameter Settings and Evaluation

We present the parameter settings of our scheme to achieve 128-bit security. We set $p = 2^{38} - 107, d = 128, k = 4, N = 2, M = 3, \tau = 5, \omega = 4, p_e = 3329, N_e = 4, K_e = 9$. For the user identity space $\mathcal{I} = \{i(X^k) \in R_{kd,p} : i \in \mathcal{B} \wedge \|i\|_1 = \omega\}$, we set $\omega = 4$. Therefore, the approximate maximum number of group members $\mathcal{N}$ that each group can accommodate in our scheme is about 2^{23}.

Recall our scheme's public key $gpk = (\mathbf{A}, \mathbf{B}, \mathbf{B}', \mathbf{u}, \mathbf{A_e}, \mathbf{b_e}) \in R_{kd,p}^{N \times (N+M)} \times R_{kd,p}^{N \times \tau N} \times R_{kd,p}^{N \times \tau N} \times R_{kd,p}^{N} \times R_{p_e}^{N_e \times K_e} \times R_{p_e}^{K_e}$. The size is $(N(N+M) + 2\tau N^2)kd \times \log_2 p + (N_e K_e + K_e)d \times \log_2 p_e \approx 127.0\text{KB}$. While   our scheme's private signing key $gsk_i = (\mathbf{s_i^1}, \mathbf{s_i^2}, \mathbf{s_i^3}) \in R_{kd,p}^{N+M} \times R_{kd,p}^{\tau N} \times R_{kd,p}^{\tau N}$, thus the size is $(N+M+2\tau N)kd \times \log_2 p \approx 59.4$ KB. For the message signature Σ, represent it as $(\zeta, \tilde{\mathbf{v}}, \mathbf{t_0}, \mathbf{t_1}, \Pi, t_s)$, then the size is $2Nkd \times \log_2 p + (N_e+1)d \times \log_2 p_e + S_\Pi + S_{t_s} \approx 93.5$ KB, where S_Π is the size of the ZKAoK Π and S_{t_s} is the size of the timestamp t_s, set to 32 bits.

Figure 4 and Fig. 5 respectively present the comparison situations of our scheme and other lattice-based schemes [6,17] in terms of public key size and signature size. It is evident from the figures that the public key size and signature size of the Schemes [6,17] increase as the number of members grows. However, the sizes of the public key and the signature in our scheme are not affected by the number of members. Hence, our scheme is more suitable for the vehicle network scenarios with a high density of vehicles.

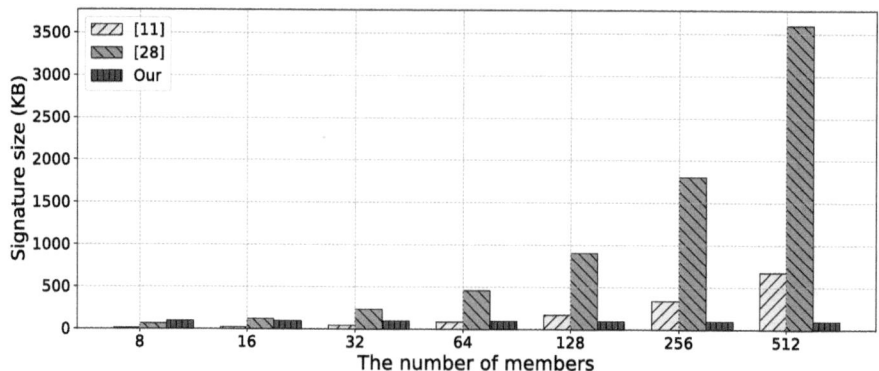

Fig. 5. Comparison of Signature sizes

8   Conclusion

We propose an efficient privacy-preserving authentication protocol, aiming to address the security and privacy challenges in VANETs. The protocol utilizes M-SIS and M-LWE problems, achieving quantum resistance while meeting characteristics such as anonymity, traceability, and replay attack resistance. Not only does this scheme effectively protect user privacy, but it also enables effective tracking of malicious behavior, balancing system security and privacy needs. Additionally, the protocol has low communication overhead. In summary, we provide a feasible and efficient solution for secure communication in VANETs.

Acknowledgements. This work is financially supported by the National Natural Science Foundation of China under Grant No. 12441102 and the self-determined research funds of CCNU from the colleges basic research and operation of MOE under Grant No. CCNU24ai010.

References

1. Biswas, S., Tatchikou, R., Dion, F.: Vehicle-to-vehicle wireless communication protocols for enhancing highway traffic safety. IEEE Commun. Mag. **44**(1), 74–82 (2006)
2. Brakerski, Z., Gentry, C., Vaikuntanathan, V.: (Leveled) fully homomorphic encryption without bootstrapping. ACM Trans. Comput. Theory (TOCT) **6**(3), 1–36 (2014)
3. Cao, Y., Xu, S., Chen, X., He, Y., Jiang, S.: A forward-secure and efficient authentication protocol through lattice-based group signature in vanets scenarios. Comput. Netw. **214**, 109149 (2022)
4. Chang, S., Qi, Y., Zhu, H., Zhao, J., Shen, X.: Footprint: detecting sybil attacks in urban vehicular networks. IEEE Trans. Parallel Distrib. Syst. **23**(6), 1103–1114 (2011)

5. Hubaux, J.P., Capkun, S., Luo, J.: The security and privacy of smart vehicles. IEEE Secur. Privacy **2**(3), 49–55 (2004)
6. Jiao, C., Xiang, X.: Anti-quantum lattice-based ring signature scheme and applications in vanets. Entropy **23**(10), 1364 (2021)
7. Kenney, J.B.: Dedicated short-range communications (dsrc) standards in the united states. Proc. IEEE **99**(7), 1162–1182 (2011)
8. Langlois, A., Stehlé, D.: Worst-case to average-case reductions for module lattices. des. Codes Crypt. **75**(3), 565–599 (2014)
9. Lin, X., Sun, X., Ho, P.H., Shen, X.: Gsis: a secure and privacy-preserving protocol for vehicular communications. IEEE Trans. Veh. Technol. **56**(6), 3442–3456 (2007)
10. Lyubashevsky, V., Nguyen, N.K., Plançon, M.: Lattice-based zero-knowledge proofs and applications: shorter, simpler, and more general. In: Annual International Cryptology Conference, pp. 71–101. Springer (2022)
11. Micciancio, D., Peikert, C.: Trapdoors for lattices: simpler, tighter, faster, smaller. In: Annual International Conference on the Theory and Applications of Cryptographic Techniques, pp. 700–718. Springer (2012)
12. Mundhe, P., Yadav, V.K., Singh, A., Verma, S., Venkatesan, S.: Ring signature-based conditional privacy-preserving authentication in vanets. Wireless Pers. Commun. **114**, 853–881 (2020)
13. Mundhe, P., Yadav, V.K., Verma, S., Venkatesan, S.: Efficient lattice-based ring signature for message authentication in vanets. IEEE Syst. J. **14**(4), 5463–5474 (2020)
14. Ng, S.C., Zhang, W., Zhang, Y., Yang, Y., Mao, G.: Analysis of access and connectivity probabilities in vehicular relay networks. IEEE J. Sel. Areas Commun. **29**(1), 140–150 (2010)
15. Shao, J., Lin, X., Lu, R., Zuo, C.: A threshold anonymous authentication protocol for vanets. IEEE Trans. Veh. Technol. **65**(3), 1711–1720 (2015)
16. Toor, Y., Muhlethaler, P., Laouiti, A., De La Fortelle, A.: Vehicle ad hoc networks: applications and related technical issues. IEEE Commun. Surv. Tutor. **10**(3), 74–88 (2008)
17. Wen, J., Bai, L., Yang, Z., Zhang, H., Wang, H., He, D.: Larrs: Lattice-based revocable ring signature and its application for vanets. IEEE Trans. Vehicular Technol. (2023)
18. Yue, X., Chen, B., Wang, X., Duan, Y., Gao, M., He, Y.: An efficient and secure anonymous authentication scheme for vanets based on the framework of group signatures. IEEE Access **6**, 62584–62600 (2018)
19. Zeng, S., Huang, Y., Liu, X.: Privacy-preserving communication for vanets with conditionally anonymous ring signature. Int. J. Network Secur. **17**(2), 135–141 (2015)
20. Zhang, L., Wu, Q., Qin, B., Domingo-Ferrer, J., Liu, B.: Practical secure and privacy-preserving scheme for value-added applications in vanets. Comput. Commun. **71**, 50–60 (2015)
21. Zhang, L., Li, C., Li, Y., Luo, Q., Zhu, R.: Group signature based privacy protection algorithm for mobile ad hoc network. In: 2017 IEEE International Conference on Information and Automation (ICIA), pp. 947–952. IEEE (2017)
22. Zhu, X., Jiang, S., Wang, L., Li, H.: Efficient privacy-preserving authentication for vehicular ad hoc networks. IEEE Trans. Veh. Technol. **63**(2), 907–919 (2013)

A Privacy-Preserving Computer-Aided Diagnosis Framework for Medical Applications Using Federated Learning and Homomorphic Encryption

Jichao Xiong[1,2], Jiageng Chen[1,2](✉), Hui Liu[1], Guangyou Zhou[1,2], Jianqun Cui[1], Junyu Lin[1,2], and Dian Jiao[1,2]

[1] School of Computer Science, and Hubei Provincial Key Laboratory of Artificial Intelligence and Smart Learning, Central China Normal University, Wuhan 430079, China
`jiageng.chen@ccnu.edu.cn`
[2] School of Central China Normal University Wollongong Joint Institute, Central China Normal University, Wuhan 430079, China

Abstract. Skin cancer is one of the most common cancers worldwide, where early detection significantly reduces mortality rates. Integrating Internet of Things (IoT) technologies into healthcare enables continuous monitoring and early diagnosis through connected devices, offering new prospects for real-time medical assessment. Advanced machine learning classifiers have shown superior performance over human experts in diagnosing pigmented skin lesions, underscoring their potential for IoT-based applications. However, training high-precision computer-aided diagnosis (CAD) systems requires extensive data, and secure data sharing poses significant challenges due to privacy concerns in IoT healthcare networks. This paper introduces an IoT-oriented skin cancer CAD system combining federated learning and homomorphic encryption to address these issues. Federated learning supports collaborative model training across distributed IoT devices, mitigating data scarcity without compromising data privacy. Homomorphic encryption ensures that patient data remains encrypted during diagnosis, enhancing security within IoT frameworks. To achieve an efficient and high-accuracy model suited for IoT, we implemented Self-Learnable Activation Functions (SLAF), optimized for homomorphic encryption and resource constraints of IoT devices. Our system, validated using the HAM10000 dataset, achieved 94.39% accuracy with dual privacy protection, enhancing both diagnostic precision and data security. Comprehensive evaluations and comparisons with state-of-the-art frameworks confirmed the effectiveness of our approach. Results demonstrate the strong potential and practicality of our IoT-based solution for real-world healthcare, providing a secure, efficient, and accurate diagnostic tool that supports the wider adoption of IoT in medical applications.

Keywords: Skin Cancer Classification · Computer Aided Diagnosis · Privacy-Preserving Machine Learning

© The Author(s), under exclusive license to Springer Nature Switzerland AG 2025
W. Meng et al. (Eds.): ADIoT 2024, LNCS 15397, pp. 90–106, 2025.
https://doi.org/10.1007/978-3-031-85593-1_6

1   Introduction

Skin cancer is one of the most commonly diagnosed cancers worldwide. Malignant skin tumors, also known as melanomas, typically begin when melanocyte cells start to grow uncontrollably [15]. Among various types of skin tumors, malignant melanoma is the most aggressive and deadly due to its high propensity to invade nearby tissues, significantly contributing to global cancer mortality rates. In the United States alone, the annual treatment cost for skin cancer exceeds 8 billion dollars, imposing a substantial economic burden on the healthcare system. Early detection of malignant melanoma is crucial for improving patient survival rates. Research indicates that if this cancer is identified in its early stages, the five-year survival rate can be as high as 99%; however, if diagnosis is delayed, the survival rate sharply decreases to 23% [8]. Therefore, timely and accurate diagnosis and intervention are essential for improving the prognosis of patients with malignant melanoma.

During skin cancer screenings, relying on visual diagnosis often leads to misdiagnoses due to the similarities between skin lesions and normal tissue. Even experienced dermatologists face significant challenges in this diagnostic task [16]. With the rapid advancement of machine learning and image recognition technologies, computer-aided diagnosis (CAD) systems have emerged as valuable assessment tools [31]. Studies have shown that state-of-the-art machine learning classifiers outperform human experts in diagnosing pigmented skin lesions, indicating their potential for greater utilization in clinical practice [28]. Therefore, developing a reliable and efficient CAD system can provide physicians with a powerful second opinion, aiding in the earlier and more accurate diagnosis of patients' conditions, thus improving treatment outcomes and increasing survival rates.

The healthcare sector is undergoing a significant transformation with the integration of IoT (Internet of Things) technologies, especially in medical diagnosis and treatment [20]. The IoT provides a network of interconnected devices, capable of gathering, processing, and transmitting patient data in real time, which offers numerous possibilities for improving healthcare systems. In this context, the combination of IoT and AI (Artificial Intelligence) is paving the way for a new paradigm called the Artificial Intelligence of Things (AIoT), where AI adds an intelligent layer to IoT, enabling smart and autonomous medical decision-making systems. This integration is instrumental in advancing the capabilities of computer-aided diagnosis systems, such as those used for skin cancer detection, by providing continuous monitoring, real-time analysis, and distributed data processing [11]. The fusion of IoT, AI, edge-fog-cloud computing, and distributed ledger technologies (DLTs) further enhances the reliability and efficiency of medical diagnostics. The IoT generates vast amounts of heterogeneous data from interconnected medical devices, while AI helps in analyzing this data to derive actionable insights. Edge-fog-cloud computing improves the scalability and responsiveness of IoT systems by ensuring that data processing happens closer to the source of data generation, reducing latency and improving real-time decision-making capabilities.

In the healthcare sector, the sensitivity of data is crucial for accurately predicting and detecting diseases [1]. However, the training of high-accuracy machine learning models requires large datasets, complicating the acquisition of medical data. The phenomenon of "data silos" in healthcare, due to the sensitivity of the data, poses a significant barrier to resource sharing. Most healthcare institutions are geographically dispersed and adhere to their administrative regulations, making them reluctant to risk violating privacy ethics or losing economic benefits by sharing patient medical data [32]. Additionally, patients face privacy leakage risks when using machine learning models for disease pre-assessment. This situation hinders the effective integration and sharing of data, limiting the application and development of machine learning technology in the medical field. Therefore, addressing the conflict between data sharing and privacy protection is crucial for advancing the use of machine learning in healthcare.

1.1   Computer Aided Diagnosis Systems for Skin Cancer

Mohammed A. Al-masni and colleagues [2] proposed a comprehensive diagnostic framework that integrates a skin lesion boundary segmentation stage with multiple skin lesion classification stages. This framework first employs a deep learning Full-Resolution Convolutional Network (FrCN) to segment the skin lesion boundaries from the entire dermoscopic image. Subsequently, convolutional neural network classifiers (such as Inception-v3, ResNet-50, Inception-ResNet-v2, and DenseNet-201) are applied to the segmented skin lesions for classification.

Jie Song and colleagues [27] proposed an innovative neural network ensemble model composed of three main parts. First, U-net is used as a segmentation network to generate masks of skin lesions, which are then used to crop the original images. Second, advanced deep convolutional neural networks (DCNN) are employed to extract features from the cropped images, incorporating squeeze-excitation blocks (SE Blocks) to emphasize and strengthen useful features. Finally, a novel neural network with local connections is constructed to integrate classification results, extract features of different categories, and individually consolidate the classification results for each category.

Additionally, Muhammad Attique Khan and colleagues [17] proposed a deep learning-based system comprising two main steps: segmentation and classification. In the segmentation step, the MASK-RCNN model is used to segment skin lesions; in the classification step, the DenseNet model is employed to classify the lesions. Furthermore, the system optimizes feature extraction through transfer learning to enhance model performance. On the other hand, Vatsala Anand and colleagues [4] proposed a hybrid model that combines U-Net and convolutional neural network models. This hybrid model first uses the U-Net model to segment skin images and then applies convolutional neural network models to perform multi-class classification on the segmented images. When simulated and analyzed using the HAM10000 dataset, this model achieved an accuracy rate of up to 97.96%. These studies indicate that integrating different deep learning techniques and models can significantly improve the accuracy of skin lesion detection and classification, providing robust technical support for clinical practice.

1.2   Privacy Preserving in Medical Diagnosis

In the United States, data used to support public health surveillance and research must be de-identified before being released to the public to address privacy concerns. Abinitha Gourabathina and colleagues introduced a game theory model that accounts for the dynamic nature of infection rates and can adaptively generate data release strategies. This model models the data release process as a two-player Stackelberg game between the data publisher and the data receiver, then seeks the optimal strategy for the publisher [12].

MedPFL is a framework designed for privacy risk analysis and mitigation in federated learning for medical images. This framework employs various levels of random noise to defend against federated learning attacks, noting that while higher noise levels can enhance privacy protection, they may still be insufficient. The article uses the example of privacy risks encountered when handling medical data in Florida to demonstrate real-world scenarios of privacy attacks on medical images across benchmark datasets, further illustrating the critical challenges of mitigating privacy risks in the healthcare domain [9].

Eid Albalawi and colleagues proposed a federated learning-based deep learning model for the automatic and accurate classification of brain tumors. This model leverages the powerful capabilities of Convolutional Neural Networks (CNN), specifically a modified version of the VGG16 architecture optimized for brain MRI images. This innovative approach not only underscores the significance of the modified VGG16 architecture in handling brain tumor classification but also highlights the crucial roles of federated learning and transfer learning in medical imaging. By incorporating transfer learning during the model training process, the model significantly enhances its classification performance, demonstrating the immense potential of this approach in improving the accuracy of medical imaging diagnostics [3].

1.3   Our Contribution

Efficient Privacy-Preserving Computer Aided Diagnosis System for Skin Cancer: This paper introduces an efficient privacy-preserving computer aided diagnosis system for skin cancer by integrating two cryptographic tools: federated learning and homomorphic encryption. To validate the effectiveness of our proposed system, we tested it using the real-world HAM10000 dermatology dataset and employed data augmentation techniques to appropriately expand and balance the dataset.

The experimental results indicate that our privacy-preserving computer-aided diagnosis system for skin cancer surpasses most similar systems in classification accuracy, achieving an accuracy of 94.39% on the HAM10000 dataset. Additionally, our system supports two privacy protection features not present in other comparable solutions. On a device equipped with a 16GB Apple M1 Pro CPU (8-core), a single diagnosis using encrypted data takes approximately one minute. This system excels not only in accuracy and efficiency but also in safeguarding patient data privacy, validating its feasibility and practicality for real-world medical diagnostics.

Solving the "Data Silos" Problem in Healthcare: Traditional machine learning methods require centralized data for training, which fails to meet the critical confidentiality needs in medical data processing. This is because patients' relevant medical data is typically stored only in the hospital databases where they receive treatment, and due to privacy protections, hospitals cannot share patients' private data. However, relying solely on the medical data held by a single hospital is insufficient for training an efficient machine learning model due to the limited data volume.

To address this issue, this paper utilizes a Horizontal Federated Learning mechanism to construct a privacy-preserving machine learning model training framework. This framework allows hospitals to "share" patient medical data while protecting patient privacy. In this framework, each hospital uses its local database of patient data to train a local model and only needs to share the parameters and update information of its local model to collaboratively train a highly accurate global model. This approach improves the model's performance and generalization ability while safeguarding privacy.

Addressing Privacy Data Leakage in Patient Prognosis Prediction: When using machine learning-based computer-aided diagnosis systems, patients need to upload their medical information for diagnosis. However, uploading plaintext data can easily lead to privacy data leaks. To solve this issue, this paper develops an integrated framework that combines homomorphic encryption with Convolutional Neural Networks (CNN) to support privacy-preserving machine learning.

Within this framework, patients do not need to upload plaintext medical data. Instead, they encrypt their medical data using homomorphic encryption before uploading it to the diagnosis system. Homomorphic encryption allows computations to be performed on encrypted data, enabling the computer-aided diagnosis system to process encrypted data and generate diagnostic results without needing decryption. This method ensures that patients' privacy data remains encrypted throughout the entire diagnostic process, significantly reducing the risk of data leakage.

2   Preliminaries

2.1   Federated Learning

Federated Learning (FL) is a distributed machine learning paradigm that enables the training of a model across multiple decentralized devices or servers holding local data samples, without exchanging them. This approach addresses the challenge of data privacy by allowing model training to occur locally, ensuring that sensitive data remains on the device and only model updates (gradients) are communicated to a central server [21].

The FL process typically involves the following steps:

– Model Initialization: A global model is initialized on the server side, and its parameters are sent to multiple local clients.
– Local Training: Each client receives the global model parameters and trains the model using its local data. After training is complete, each client sends the updated model parameters back to the server.
– Global Aggregation: The server receives the model parameters from each client and updates the global model parameters by performing a weighted average of these parameters.
– Iteration: The updated global model is redistributed to each device, and the entire process is repeated until convergence is achieved.

Federated learning offers a powerful and privacy-preserving paradigm for advancement. By distributing model training tasks across multiple clients while keeping the data processing localized, this approach effectively addresses data privacy concerns. Federated learning not only avoids the privacy risks associated with data sharing but also enhances the model's robustness and generalization capability by leveraging diverse datasets.

2.2 Fully Homomorphic Encryption

The concept of Fully Homomorphic Encryption (FHE) was first introduced by Rivest et al. in 1978, originally termed as privacy homomorphisms [24]. FHE is a special type of encryption method that allows computations to be performed on encrypted data without decrypting it. This means that even while data remains in an encrypted state, operations such as addition and multiplication can be carried out, and the decrypted result will be the same as if the operations had been performed on the unencrypted data. This characteristic of homomorphic encryption endows it with immense potential and application prospects in privacy protection, especially offering robust privacy mechanisms in fields such as machine learning, cloud computing, data processing, and data sharing.

The CKKS homomorphic encryption scheme, based on the Ring Learning With Errors (RLWE) problem, is a type of approximate arithmetic homomorphic encryption method [7]. Unlike the BGV [6] and FV [10] schemes, the plaintext space in CKKS consists of complex polynomials. CKKS uses complex space as its message space, supporting approximate addition and multiplication operations on encrypted data. It introduces a new scaling mechanism to manage the scale of the plaintext.

In this study, we used the CKKS levelled-FHE implemented in the TenSEAL [5] python wrapper library provided by Microsoft SEAL [26] in all our experiments.

3 Designing an Efficient Privacy-Preserving Computer Aided Diagnostic System for Skin Cancer

This paper proposes a comprehensive framework that combines federated learning and homomorphic encryption to develop a computer-aided diagnosis system, specifically designed for integration into IoT-based healthcare environments. The primary objective of this framework is to deliver high-accuracy skin disease classification, facilitating medical professionals in diagnosing conditions while ensuring data privacy and security. Key features of this system include Distributed Data Training and Encrypted Data Diagnosis, optimized for IoT deployments.

Distributed Data Training enables hospitals and medical facilities within an IoT network to collaboratively develop robust machine learning models for disease classification without the need to share confidential patient data. This feature not only addresses the challenge of limited data availability within individual institutions but also aligns with the decentralized nature of IoT infrastructures, where devices and data sources are geographically dispersed. By leveraging this capability, the framework supports real-time data processing and model training across connected devices, enhancing the scalability and responsiveness of the diagnostic system.

Encrypted Data Diagnosis allows patients to utilize the computer-aided diagnosis system while maintaining the confidentiality of their medical data. Within an IoT context, where data is continuously generated and transmitted across devices, this feature ensures that patients' personal data remains encrypted throughout the diagnostic process. The use of homomorphic encryption facilitates computations on encrypted data, allowing for secure diagnosis without exposing sensitive information. This mechanism significantly enhances data security and minimizes privacy risks, a critical requirement for IoT-driven healthcare applications.

3.1 Constructing a Privacy-Preserving Machine Learning Model Supporting Distributed Data Training

Wahab et al. utilized federated learning technology to construct a privacy-aware framework for anomaly detection in wireless capsule endoscopy. The study results indicated that the decentralized model based on federated learning improved accuracy by 10–12% compared to the best accuracy of a centralized model [30]. To address the issue of limited medical data within individual hospitals, which hinders the training of high-accuracy machine learning models, and to overcome the restriction on sharing private medical data between hospitals, this paper introduces a horizontal federated learning mechanism to ensure the confidentiality of training the computer aided diagnosis system. In this framework, hospital nodes are required to use the same network model and data with the same feature dimensions.

Assume there are K independent hospitals, each holding a portion of patients' skin examination data and diagnostic results. The distributed data training process for a privacy-preserving computer-aided diagnostic system for skin cancer is as follows:

1. Model Download: Each hospital node downloads the unified network model and input data requirements from the cloud server.
2. Local Training: Each hospital trains the local model using its local database.

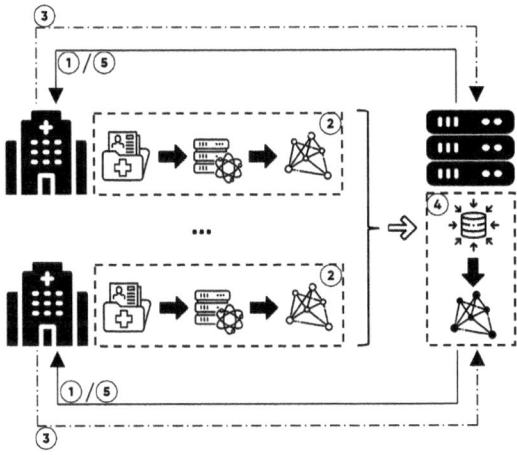

Fig. 1. Distributed Data Training Network Model Description Diagram

3. Parameter Upload: Each hospital uploads the parameters of the locally trained model to the cloud server.
4. Global Model Update: The cloud server receives the model parameters from each hospital and performs a weighted average operation on these local model parameters to update the global model parameters.
5. Parameter Distribution: Each hospital node downloads the updated global model parameters.

By iterating steps 2 to 4 multiple times, the system can fully utilize the data resources of each hospital to build a high-accuracy skin cancer diagnosis model while ensuring data privacy. Figure 1 illustrates the privacy-preserving machine learning model training process for skin cancer diagnosis.

3.2   Constructing a Privacy-Preserving Machine Learning Model Supporting Encrypted Data Diagnosis

In traditional computer aided diagnosis systems, doctors or patients need to upload the patient's relevant medical data to assist in diagnosis, which can easily lead to the leakage of patient privacy data. In the framework proposed in this paper, each hospital downloads the updated high-accuracy global model from the cloud server and deploys it on their local server. Subsequently, the CKKS homomorphic encryption scheme is used to transform the network model into a privacy-preserving network model that supports computations on encrypted data.

Under this privacy-preserving network model, patients need to register a public-private key pair on the hospital server during their visit. After obtaining dermoscopic examination data, the patient first encrypts their medical data using the public key and then uploads the encrypted data to the hospital server for auxiliary diagnostic prediction. The hospital server processes the patient's data while it remains encrypted and computes the prediction result. The prediction result, still in encrypted form, is then returned to the patient. Since the private key is held only by the patient, the confidentiality of the prediction result is ensured. Meanwhile, the hospital server continuously processes data in its encrypted state, effectively safeguarding patient data privacy.

However, homomorphic encryption schemes typically only support addition and multiplication operations, making the direct implementation of traditional nonlinear activation functions challenging [22]. To overcome these limitations and address the issue of reduced model classification accuracy caused by traditional low-degree polynomial functions, we introduce the Self-Learnable Activation Functions (SLAF) [13]. SLAF is a generalized activation function, with the general form $a_0x^0 + a_1x^1 + a_2x^2 + ... + a_nx^n$, where $a_0, a_1, a_2, ..., a_n$ are parameters learned during training. SLAF retains the properties of polynomial functions, making it easily implementable within homomorphic encryption schemes. Consequently, SLAF can be adjusted for specific datasets and network structures, effectively improving the accuracy of traditionally homomorphic encryption-friendly networks.

The privacy-preserving network model used for diagnostic prediction is illustrated in Fig. 2.

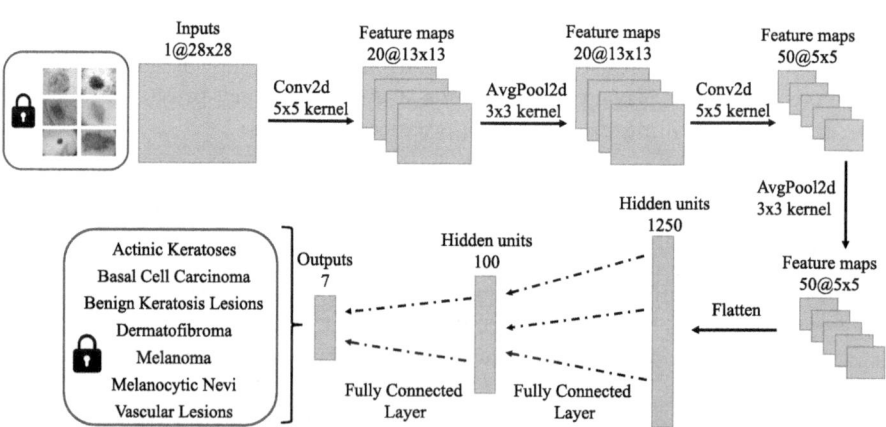

Fig. 2. Diagnostic Neural Network Architecture for Encrypted Data

4   Experimental Evaluation

To validate the effectiveness and practicality of our proposed privacy-preserving computer-aided diagnosis system, we conducted experiments using the publicly available HAM10000 dataset. The classification performance was assessed based on the specific accuracy for each of the seven types of skin diseases, as well as the overall classification accuracy of the model. To ensure comprehensive evaluation, we compared our system with other state-of-the-art skin cancer diagnostic models.

HAM10000: The HAM10000 dataset [29] is a robust collection of 10,015 dermoscopic images, assembled from diverse sources to address the typical limitations in dataset diversity and size that impede machine learning model development. It includes images representing seven clinically significant categories of skin lesions: Melanoma (MEL), Vascular Lesions (VASC), Benign Keratosis Lesions (BKL), Dermatofibroma (DF), Melanocytic Nevi (NV), Basal Cell Carcinoma (BCC), and Actinic Keratoses (AKIEC). This dataset serves as a reliable benchmark for training and validating automated diagnostic algorithms, supporting applications that mimic real-world clinical settings.

A significant proportion of the dataset (over 50%) has been verified by histopathology, while the remaining cases have undergone validation through follow-up, expert consensus, or in vivo confocal microscopy. The diverse origins of the images, sourced from different medical institutions and spanning various acquisition methods, contribute to the dataset's strength. This diversity is crucial for training models that generalize well across different imaging conditions and patient demographics, aligning perfectly with IoT-based healthcare applications where data heterogeneity is common.

In IoT-enabled diagnostic systems, such a dataset is invaluable. The HAM10000 dataset's diversity supports the development of machine learning models that can seamlessly integrate into IoT infrastructures, facilitating the deployment of distributed diagnostic tools capable of real-time, remote analysis. This capability is essential for enhancing access to high-quality dermatological diagnostics in various clinical and non-clinical settings, leveraging IoT to connect medical devices, process patient data securely, and deliver diagnostic results efficiently.

4.1   Data Augmentation

Effective training of machine learning classification networks requires a large amount of training data. However, automatic diagnosis of skin lesions is still constrained by the limited number of images and the lack of ground-truth annotations in existing dermoscopic image databases. In the HAM10000 dataset, there is a significant disparity in the number of images across different categories. To address this issue, we performed data augmentation on the HAM10000 dataset to increase the number of training images and avoid overfitting that might occur with a small amount of training data.

Specifically, we achieved data augmentation by rotating the original images at specific angles. To address the disparity in the number of images among different categories, we applied different numbers of rotation operations, resulting in a more balanced dataset. The number of images in each category before and after data augmentation is shown in Table 1. After data augmentation, we resized the images to a standard size of $28 \times 28 \times 1$ to meet the input requirements of the model. Subsequently, we randomly divided the dataset in an 8:2 ratio, with 77,484 images used for training and 19,372 images used for testing.

Table 1. Comparison of accuracy and time overhead of different models

	Before Augmentation	After Augmentation
AKIEC	327	13407
BCC	514	13878
BKL	1099	14287
DF	115	13915
NV	6705	13410
VASC	142	13490
MEL	1113	14469
Total	10015	96856

4.2 Local Training Iterations and Global Aggregation Rounds Parameter Settings

To optimize the model's parameter settings, we first evaluated the impact of different local training cycles (local epoch) on model accuracy under the condition of a single communication between each hospital node and the cloud server. Specifically, we experimented with local epoch values of 1, 10, 30, 60, 90 to 300, and recorded the model accuracy corresponding to each local epoch value. The results are shown in Fig. 3.

The experimental results indicated that setting the local epoch to 150 yielded the highest model accuracy. Therefore, with the local epoch fixed at 150, we further investigated the impact of the communication frequency between hospital nodes and the cloud server on model accuracy. We experimented with different communication frequencies, ranging from 1 to 7 times. The results are shown in Fig. 4.

4.3 Model Performance Metrics

To comprehensively evaluate the performance of the proposed model under optimal parameter settings, we tested the specific classification accuracy for the seven

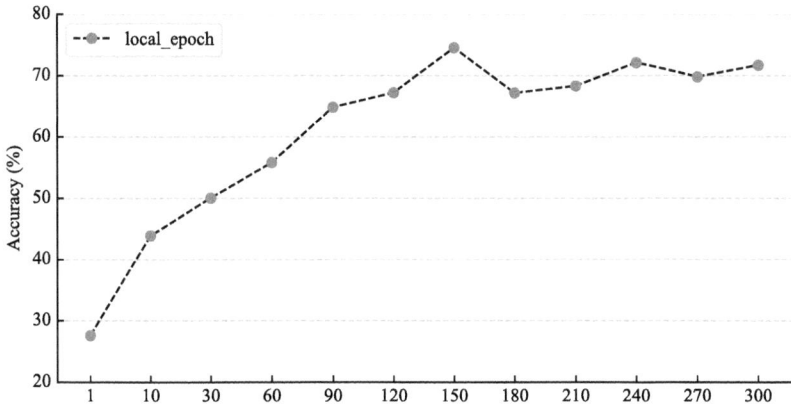

Fig. 3. The Impact of Different Local Epochs on Model Accuracy

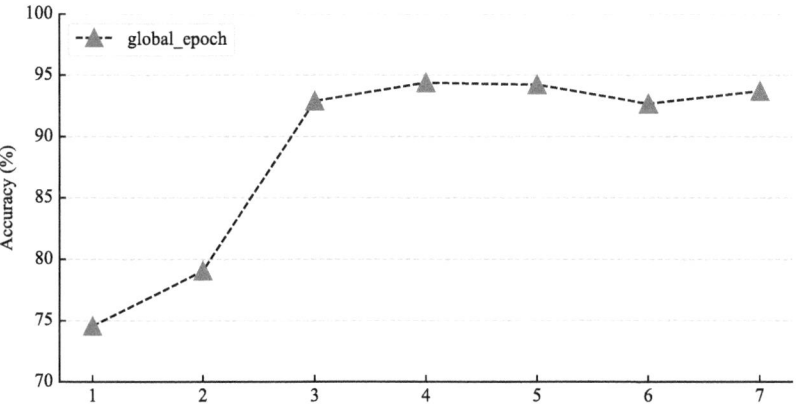

Fig. 4. The Impact of Different Global Epochs on Model Accuracy

types of skin diseases as well as the overall classification accuracy of the model. Figure 5 illustrates the model's specific performance across different disease classifications. These evaluation results provide an in-depth understanding of the model's performance across various dimensions, validating the effectiveness and efficiency of the proposed model in practical applications.

4.4 Experimental Comparisons

To verify the practicality and effectiveness of the proposed method, we compared it with existing similar skin cancer computer aided diagnosis systems. Table 2 provides a detailed comparison of the functionality and accuracy between our proposed solution and similar existing systems.

The experimental results indicate that our method exhibits superior accuracy compared to most existing skin cancer computer-aided diagnosis systems, with

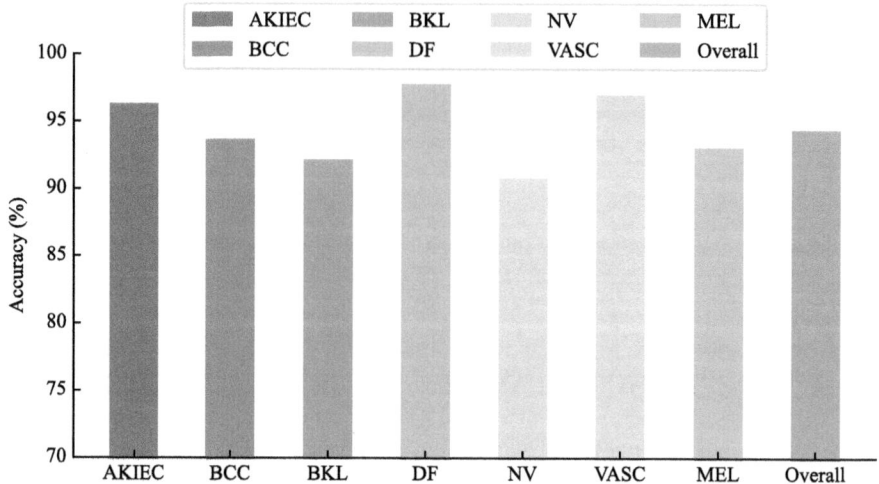

Fig. 5. Individual Classifications and overall Accuracy under Optimal Parameters

Table 2. Comparison of Functionality and Accuracy with Other Skin Cancer Computer-Aided Diagnosis Systems

Scheme	Skin Disease Segmentation	Skin Lesion Classification	Distributed Data Training	Encrypted Data Diagnosis	Accuracy
[2]	●	●	○	○	88.70%
[14]	○	●	○	○	85.80%
[25]	○	●	○	○	83.15%
[23]	○	●	○	○	92.90%
[17]	●	●	○	○	88.50%
[18]	●	●	○	○	90.67%
[4]	●	●	○	○	97.96%
Ours	○	●	●	●	94.39%*

"*" represents the accuracy of classification diagnosis using encrypted data.

performance slightly lower than the approach proposed by Anand et al. [4]. We attribute this slight difference to our method's use of entire dermoscopic images for diagnostic classification, rather than employing Anand et al.'s strategy of first using a U-Net network to segment skin lesions before classification.

Effective lesion segmentation can enhance the accuracy of machine learning models in classifying diseased tissue but might also risk missing some lesion details due to segmentation errors. Additionally, malignant melanomas tend to spread to adjacent tissues [19], making it important to retain details of the surrounding tissue for diagnostic classification.

Besides demonstrating competitive accuracy, our method also excels in protecting medical data privacy. The proposed solution effectively addresses the issue of insufficient data in individual hospitals for training high-precision models and the inability of hospitals to share data due to policy and privacy constraints. It also supports patients in using encrypted data for lesion classification diagnosis.

Overall, our computer-aided diagnosis system combines high accuracy with privacy protection, showcasing significant usability and practicality. Table 2 provides a detailed comparison of the functionality and accuracy of our method against similar existing solutions, further validating its practicality and effectiveness.

5   Discussion

While our proposed system demonstrates significant potential for privacy-preserving diagnostics in IoT-based healthcare, it is essential to recognize its limitations. One notable challenge is the computational overhead introduced by integrating privacy-preserving mechanisms such as fully homomorphic encryption (FHE). Although FHE ensures that patient data remains encrypted throughout the diagnostic process, its high computational requirements can lead to increased latency and energy consumption, which may strain IoT devices with limited processing power. This trade-off between privacy protection and computational efficiency needs to be carefully managed, especially in resource-constrained IoT environments where real-time performance is critical.

Another limitation lies in the reliance on a trusted cloud server within the federated learning (FL) framework. While FL helps in decentralizing the training process and mitigating the risk of data centralization, the aggregation of local model updates at a cloud server necessitates trust in the server's integrity and security. Addressing this reliance by incorporating secure aggregation techniques and improving the robustness of the server architecture could further enhance the trustworthiness of the system in IoT deployments.

Moreover, the combination of homomorphic encryption and deep neural network architectures imposes restrictions on model complexity. Current implementations of FHE typically support a limited set of operations, which constrains the depth of the neural networks and the types of activation functions that can be employed. These limitations may affect the overall performance of the model, especially when compared to unencrypted counterparts that can leverage more complex structures and non-linear activation functions. Exploring advancements in efficient encrypted computation and adaptive encryption-friendly network designs will be crucial for future improvements.

For future work, we plan to extend our system by incorporating secure aggregation methods into the federated learning process to bolster data protection during model updates. Additionally, enhancing the homomorphic encryption scheme to support a broader range of neural network operations without compromising computational efficiency will be a focus. These enhancements aim to further adapt our system to IoT healthcare scenarios, enabling secure, efficient, and scalable diagnostic solutions that maintain high levels of privacy Preserving.

6   Conclusion

In this study, we introduced a novel computer-aided diagnosis system that integrates federated learning and fully homomorphic encryption, merging advanced cryptographic techniques with machine learning to create a robust, privacy-preserving solution suitable for IoT-based healthcare environments. Federated learning effectively addresses the challenge of insufficient data in individual hospitals, facilitating the collaborative training of high-accuracy models without compromising data privacy or requiring centralized data sharing. This capability is particularly valuable in IoT settings, where medical data is often distributed across multiple connected devices and institutions.

Fully homomorphic encryption complements this framework by enabling diagnostic computations on encrypted data, ensuring that sensitive patient information remains secure throughout the process. This feature is crucial in IoT-driven healthcare scenarios, where data privacy and secure transmission across networks are of paramount importance.

We conducted extensive experiments using the HAM10000 dataset to evaluate the performance of our proposed system. To address dataset imbalances, we employed targeted data augmentation techniques, such as image rotation at specific angles, to balance the distribution across various lesion categories. Additionally, the implementation of Self-Learnable Activation Functions (SLAF), optimized for homomorphic encryption compatibility, further enhanced the performance of our model, demonstrating its adaptability and effectiveness in handling complex medical data.

Compared to other skin cancer diagnostic systems, our proposed solution not only achieves high accuracy but also incorporates dual privacy protection mechanisms, making it highly applicable and secure for IoT-based medical diagnostics. These features enable real-time, decentralized analysis, contributing to the broader adoption of IoT in healthcare by providing reliable, privacy-preserving, and efficient diagnostic tools that improve patient care while maintaining stringent data security.

Acknowledgments. This work is financially supported by self-determined research funds of CCNU from the colleges' basic research and operation of MOE CCNU24ai010, CCNU22JC001, and the National Natural Science Foundation of China under Grant No.62272189.

Declaration of Competing Interest. This work was financially supported by the National Natural Science Foundation of China under Grant Nos. 12441102 and 62272189, as well as by the self-determined research funds of CCNU from the colleges' basic research and operation of MOE under Grant No. CCNU24ai010.

References

1. Ahmed, S.T., et al.: Towards blockchain based federated learning in categorizing healthcare monitoring devices on artificial intelligence of medical things investigative framework. BMC Med. Imaging **24**(1), 105 (2024)
2. Al-Masni, M.A., Kim, D.H., Kim, T.S.: Multiple skin lesions diagnostics via integrated deep convolutional networks for segmentation and classification. Comput. Methods Programs Biomed. **190**, 105351 (2020)

3. Albalawi, E., et al.: Integrated approach of federated learning with transfer learning for classification and diagnosis of brain tumor. BMC Med. Imaging **24**(1), 110 (2024)
4. Anand, V., Gupta, S., Koundal, D., Singh, K.: Fusion of u-net and CNN model for segmentation and classification of skin lesion from dermoscopy images. Expert Syst. Appl. **213**, 119230 (2023)
5. Benaissa, A., Retiat, B., Cebere, B., Belfedhal, A.E.: Tenseal: a library for encrypted tensor operations using homomorphic encryption. arXiv preprint arXiv:2104.03152 (2021)
6. Brakerski, Z., Gentry, C., Vaikuntanathan, V.: (leveled) fully homomorphic encryption without bootstrapping. ACM Trans. Comput. Theory (TOCT) **6**(3), 1–36 (2014)
7. Cheon, J.H., Kim, A., Kim, M., Song, Y.: Homomorphic encryption for arithmetic of approximate numbers. In: Takagi, T., Peyrin, T. (eds.) ASIACRYPT 2017. LNCS, vol. 10624, pp. 409–437. Springer, Cham (2017). https://doi.org/10.1007/978-3-319-70694-8_15
8. Codella, N., et al.: Skin lesion analysis toward melanoma detection 2018: A challenge hosted by the international skin imaging collaboration (isic). arXiv preprint arXiv:1902.03368 (2019)
9. Das, B.C., Amini, M.H., Wu, Y.: Privacy risks analysis and mitigation in federated learning for medical images. In: 2023 IEEE International Conference on Bioinformatics and Biomedicine (BIBM), pp. 1870–1873. IEEE (2023)
10. Fan, J., Vercauteren, F.: Somewhat practical fully homomorphic encryption. Cryptology ePrint Archive (2012)
11. Firouzi, F., et al.: Fusion of iot, ai, edge-fog-cloud, and blockchain: challenges, solutions, and a case study in healthcare and medicine. IEEE Internet Things J. **10**(5), 3686–3705 (2022)
12. Gourabathina, A., Wan, Z., Brown, J.T., Yan, C., Malin, B.A.: Privacy-preserving publishing of individual-level pandemic data based on a game theoretic model. In: 2022 IEEE International Conference on Bioinformatics and Biomedicine (BIBM), pp. 961–968. IEEE (2022)
13. Goyal, M., Goyal, R., Lall, B.: Improved polynomial neural networks with normalised activations. In: 2020 International Joint Conference on Neural Networks (IJCNN), pp. 1–8. IEEE (2020)
14. Huang, H.W., Hsu, B.W.Y., Lee, C.H., Tseng, V.S.: Development of a light-weight deep learning model for cloud applications and remote diagnosis of skin cancers. J. Dermatol. **48**(3), 310–316 (2021)
15. Ilic, M., Ilic, I.: Epidemiology of pancreatic cancer. World J. Gastroenterol. **22**(44), 9694 (2016)
16. Jones, O., et al.: Dermoscopy for melanoma detection and triage in primary care: a systematic review. BMJ Open **9**(8), e027529 (2019)
17. Khan, M.A., Akram, T., Zhang, Y.D., Sharif, M.: Attributes based skin lesion detection and recognition: A mask rcnn and transfer learning-based deep learning framework. Pattern Recogn. Lett. **143**, 58–66 (2021)
18. Khan, M.A., Sharif, M., Akram, T., Damaševičius, R., Maskeliūnas, R.: Skin lesion segmentation and multiclass classification using deep learning features and improved moth flame optimization. Diagnostics **11**(5), 811 (2021)
19. Kumar, A., Vatsa, A.: Untangling classification methods for melanoma skin cancer. Front. Big Data **5**, 848614 (2022)
20. Lu, Z.x., et al.: Application of ai and iot in clinical medicine: summary and challenges. Current Med. Sci. **41**(6), 1134–1150 (2021)

21. McMahan, B., Moore, E., Ramage, D., Hampson, S., y Arcas, B.A.: Communication-efficient learning of deep networks from decentralized data. In: Artificial Intelligence and Statistics, pp. 1273–1282. PMLR (2017)
22. Nguyen, D.T.K., Duong, D.H., Susilo, W., Chow, Y.W., Ta, T.A.: Hefun: homomorphic encryption for unconstrained secure neural network inference. Future Internet **15**(12), 407 (2023)
23. Polat, K., Koc, K.O.: Detection of skin diseases from dermoscopy image using the combination of convolutional neural network and one-versus-all. J. Artif. Intell. Syst. **2**(1), 80–97 (2020)
24. Rivest, R.L., Adleman, L., Dertouzos, M.L., et al.: On data banks and privacy homomorphisms. Foundations Secure Comput. **4**(11), 169–180 (1978)
25. Salian, A.C., Vaze, S., Singh, P., Shaikh, G.N., Chapaneri, S., Jayaswal, D.: Skin lesion classification using deep learning architectures. In: 2020 3rd International conference on communication system, computing and IT applications (CSCITA), pp. 168–173. IEEE (2020)
26. Microsoft SEAL (release 3.7). https://github.com/Microsoft/SEAL (Sep 2021), microsoft Research, Redmond, WA
27. Song, J., Li, J., Ma, S., Tang, J., Guo, F.: Melanoma classification in dermoscopy images via ensemble learning on deep neural network. In: 2020 IEEE International Conference on Bioinformatics and Biomedicine (BIBM), pp. 751–756. IEEE (2020)
28. Tschandl, P., Codella, N., Akay, B.N., Argenziano, G., Braun, R.P., Cabo, H., Gutman, D., Halpern, A., Helba, B., Hofmann-Wellenhof, R., et al.: Comparison of the accuracy of human readers versus machine-learning algorithms for pigmented skin lesion classification: an open, web-based, international, diagnostic study. Lancet Oncol. **20**(7), 938–947 (2019)
29. Tschandl, P., Rosendahl, C., Kittler, H.: The ham10000 dataset, a large collection of multi-source dermatoscopic images of common pigmented skin lesions. Sci. Data **5**(1), 1–9 (2018)
30. Wahab, H., Mehmood, I., Ugail, H., Del Ser, J., Muhammad, K.: Federated deep learning for wireless capsule endoscopy analysis: Enabling collaboration across multiple data centers for robust learning of diverse pathologies. Futur. Gener. Comput. Syst. **152**, 361–371 (2024)
31. Zanddizari, H., Nguyen, N., Zeinali, B., Chang, J.M.: A new preprocessing approach to improve the performance of cnn-based skin lesion classification. Med. Biol. Eng. Comput. **59**(5), 1123–1131 (2021)
32. Zhang, L., Xu, J., Vijayakumar, P., Sharma, P.K., Ghosh, U.: Homomorphic encryption-based privacy-preserving federated learning in iot-enabled healthcare system. IEEE Trans. Network Sci. Eng. (2022)

GCFuzz: An Intelligent Method for Generating IoT Protocols Test Cases Using GAN with CVAE

Ming Zhong[1], Zisheng Zeng[1], Yijia Guo[1], Dandan Zhao[1], Bo Zhang[3], Shenghong Li[4], Hao Peng[1,2(✉)], and Zhiguo Ding[1(✉)]

[1] School of Computer Science and Technology, Zhejiang Normal University, Jinhua, Zhejiang, China
{zhongming,zeng.zs,de3mond,ddzhao,hpeng,dzg}@zjnu.edu.cn

[2] Key Laboratory of Intelligent Educational Technology and Application in Zhejiang Normal University, Jinhua 321004, China

[3] School of Cyber Science and Engineering, Shanghai Jiao Tong University, Shanghai, China
bozhangiit@sjtu.edu.cn

[4] School of Electronic Information and Electrical Engineering, Shanghai Jiao Tong University, Shanghai 200240, China
shli@sjtu.edu.cn

Abstract. The importance of Internet of Things (IoT) systems security cannot be ignored, particularly in the realm of communication. IoT protocols serve as standards for communication and interaction among devices in IoT environments. This paper enhances IoT security by identifying and exposing protocol vulnerabilities through fuzzing, a crucial method for discovering security flaws. Traditional generation-based fuzzers require reverse engineering to understand protocol grammar and generate test cases. In this paper, we propose an intelligent method called GCFuzz, which generates high-quality test cases without prior knowledge of the protocol. To understand different classes of protocol grammar, GCFuzz applies Generative Adversarial Networks (GAN) with Conditional Variational Autoencoders (CVAE) to train a generative model on various classes of protocol messages. After training, the model generates fake but credible messages. Additionally, to accurately cluster real protocol messages, we devise a novel clustering method based on a keyword. Results show that, compared to GANFuzz, Peach, BooFuzz, and Sulley, the test cases generated by GCFuzz have stronger targeting and less redundancy, while GCFuzz has higher accurate testing efficiency.

Keywords: IoT security · IoT protocols · Fuzzing · Generative Adversarial Network · Conditional Variational Autoencoders

© The Author(s), under exclusive license to Springer Nature Switzerland AG 2025
W. Meng et al. (Eds.): ADIoT 2024, LNCS 15397, pp. 107–125, 2025.
https://doi.org/10.1007/978-3-031-85593-1_7

1   Introduction

IoT protocols [1] are a set of specifications and standards used to facilitate communication and data exchange between devices in IoT [25] environments. They are typically designed to support communication requirements in fields such as Industrial Automation [4], Intelligent Environmental Monitoring [2], and Intelligent Manufacturing [32]. Common IoT protocols include Modbus-TCP [29], EtherCAT [12], MQTT [26], etc. Most attacks against IoT systems focus on security vulnerabilities in IoT protocols. Once hackers exploit vulnerabilities within these protocols to spread viruses, they can launch a remote attack without access to a physical host and threaten the IoT device's security, resulting in catastrophic consequences. For example, the Stuxnet attack [3] that occurred at Iran's nuclear facility in 2010 invaded the nuclear facility, causing serious equipment failures and damaging the facility's production capacity. Therefore, it is crucial to promptly detect and patch security vulnerabilities in implementations of IoT protocols. Fuzzing [17,18], an automated software testing technique, has become one of the most effective techniques for detecting vulnerabilities in real-world software. Although traditional generation-based protocol fuzzers have been widely adopted and detected many vulnerabilities, they still suffer from the following problems.

First, these fuzzers require a significant amount of prior knowledge, such as the formats of protocol messages (including field types, sizes, and valid value ranges, etc.). Furthermore, reasonable mutation strategies should be designed in these fuzzers. Meeting these preconditions will consume a large amount of manpower and time. Lacking this prior knowledge, the effectiveness of these fuzzers will be seriously affected. Second, despite the capability of these fuzzers to generate test cases being worthy of affirmation, a lot of non-standard test cases are generated simultaneously. Injecting all these test cases into the system under test (SUT) will waste a lot of computational resources, because the SUT has to process these invalid inputs, which may lead to unnecessary resource consumption and performance degradation. Third, the process of generating test cases with these fuzzers is specific to a particular protocol. Parameters and mutation strategies need to be adjusted for different protocol tests, due to varying prior knowledge, which results in poor portability of these fuzzers and makes it difficult to apply them to other protocols or new testing scenarios. Consequently, the complexity and cost of testing work increase correspondingly.

To address above mentioned drawbacks, we propose an intelligent fuzzer for IoT protocols, called GCFuzz (GAN [6] with CVAE [31] Fuzz). The framework of GCFuzz is shown in Fig. 1.

GCFuzz first captures real network traffic and preprocesses it. Then, GCFuzz clusters the preprocessed messages using a keyword to obtain a training dataset. Subsequently, GCFuzz employs the dataset for model training. Before training, the message sequences are converted into numerical representations ranging from 0 to 16 and spliced with corresponding labels, which are transformed into one-hot encoding, as input to the model. After training, the model parameters are loaded, and test cases are generated. These test cases are sent to SUT, han-

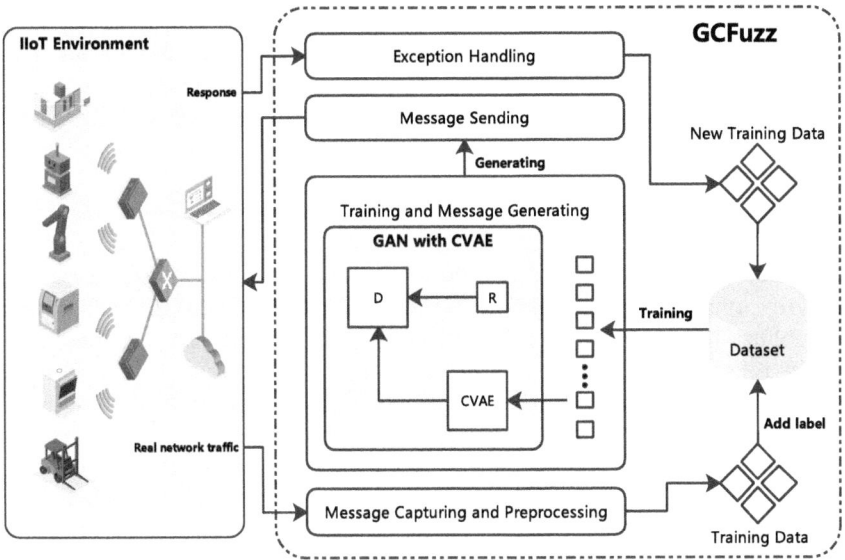

Fig. 1. GCFuzz framework

dling exceptional situations while recording test cases and abnormal information. Finally, exceptional test cases are iteratively used to train the model further for subsequent fuzzing.

Different from traditional generation-based protocol fuzzers, before using our method for fuzzing, it is necessary to cluster real protocol messages. The purpose is to allow the model to accurately learn the protocol grammar, because messages with similar structures are in the same category. Although, GANFuzz [11] proposed a clustering method based on the character set, clustering messages by applying the k-means clustering algorithm. This method has a limitation: if the length of a message far exceeds others, it is likely to be clustered into a separate category, which is not an ideal clustering result. In contrast, we propose a clustering method based on a keyword, which can accurately cluster these messages. In summary, our main contributions are as follows:

- We propose a novel clustering method for protocol messages. This method can accurately cluster protocol messages without prior knowledge.
- We design a fuzzer called GCFuzz for automated testing IoT protocols. GCFuzz accurately infers the format of the protocol message based on network traffic and generates test cases that conform to the protocol grammar.
- We evaluate GCFuzz on a real IoT protocol. The results demonstrate that, the test cases generated by GCFuzz have stronger targeting and less redundancy, while GCFuzz has higher accurate testing efficiency.

The remainder of this paper is organized as follows. In Sect. 2, we introduce the research background of this paper. Section 3 presents our proposed test case generation method. In Sect. 4, we evaluate our approach by testing Modbus-TCP and compare it with other fuzzers. In Sect. 5, we introduce some related work on fuzzing. Finally, a brief conclusion and future work are given in Sect. 6.

2   Background

2.1   Protocol Fuzzing

Fuzzing has been prevalent for over 20 years. Because of its effectiveness, fuzzing has many applications in the field of network protocol testing, such as Peach [7], BooFuzz [13], and Sulley [21], etc. During the testing process, these fuzzers play the role of a client, continuously generating packets and accurately sending them to the SUT. However, these methods have some limitations. First, these fuzzers require users to manually modify configuration parameters, mutation strategies, and provide the format of the protocol messages in order to generate high-quality test cases when testing IoT protocols. For example, when fuzzing Modbus-TCP [29] using Peach [7], we need to provide the format of Modbus-TCP packets. Figure 2 depicts the Modbus-TCP packets in the format of Peach Pit [30], the XML configuration file for the protocol fuzzer Peach. Lines 3–6 show the format of Modbus-TCP packets, including field name, size, type, etc. This process consumes a large amount of manpower and time. Second, despite manual configuration, these fuzzers still generate a lot of invalid test cases during fuzzing, leading to unnecessary resource consumption and performance degradation. For example, not all values within the entire range of fields in Modbus-TCP packets are valid. Although Peach understands the format of Modbus-TCP packets, it generates all possible values within these fields' ranges, resulting in a lot of invalid test cases. Third, different protocols have different grammars, necessitating specific configurations for each protocol when using these fuzzers. For example, testing the Modbus-TCP protocol with Peach requires writing an XML configuration file, as shown in Fig. 2. If other protocols need to be tested, new configuration files must be created. This situation results in poor portability of these fuzzers, making it challenging to apply them to other protocols or new testing scenarios. Consequently, the complexity and cost of testing increase.

2.2   Generative Adversarial Network (GAN)

GAN [6] proposed by Goodfellow et al., is a deep neural network architecture that has been widely used for generating realistic data such as images, videos, and text, demonstrating remarkable effectiveness. A GAN consists of a generative model (generator) and a discriminative model (discriminator) as shown in Fig. 3.

The goal of the generator is to deceive the discriminator into believing that the data it creates comes from the true data distribution, while the discriminator's goal is to distinguish between synthetic and real data. To achieve this, the generator receives random noise sampled from a prior distribution (such as a Gaussian distribution) to estimate a generating function that approximates the

```
1    <DataModel name="body">
2       <Block>
3          <Number name="unit_identifier" size="8" value="01" valueType="hex"
                signed="false" constraint="int(value) &lt; 5 and int(value) &gt; 0"/>
4          <Number name="function_code" size="8"/>
5          <Number name="reference_number" size="16" value="01 00" valueType="hex"
                signed="false" token="true"/>
6          <Number name="data" size="16" value="11 45" valueType="hex"
                signed="false"/>
7       </Block>
8       ...
9    </DataModel>
```

Fig. 2. A model of Modbus-TCP packets as Peach Pit

true data distribution. The discriminator is trained to maximize the probability of correctly classifying generated and real data. Conversely, the generator is trained to maximize the probability of the generated data being misclassified. Through this adversarial training approach, we can estimate a generative model that covers the real-world data distribution. In this work, we use GAN as the foundational architecture for generating protocol messages.

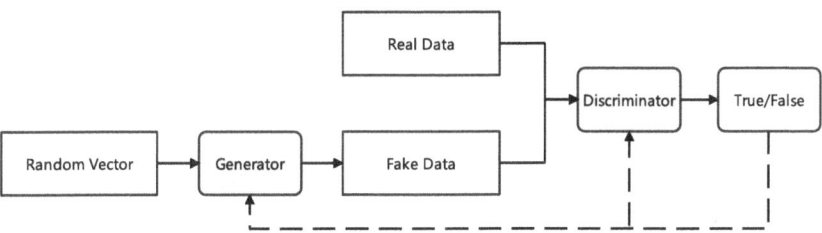

Fig. 3. The structure of GAN

2.3   Conditional Variational Autoencoders (CVAE)

CVAE [31] is a deep neural network architecture that has been widely used for generating realistic data based on certain input information (such as labels or additional features) and has demonstrated significant effectiveness. CVAE consists of an encoder and a decoder, as shown in Fig. 4, where the encoder maps the input data and conditional information to a latent space, and the decoder reconstructs the data from the latent representation and conditional information.

The goal of the encoder is to approximate the posterior distribution of the latent variables given the input data and conditional information. To achieve this, the encoder receives the input data and conditional information and outputs the parameters of the posterior distribution (mean and variance). On the other hand,

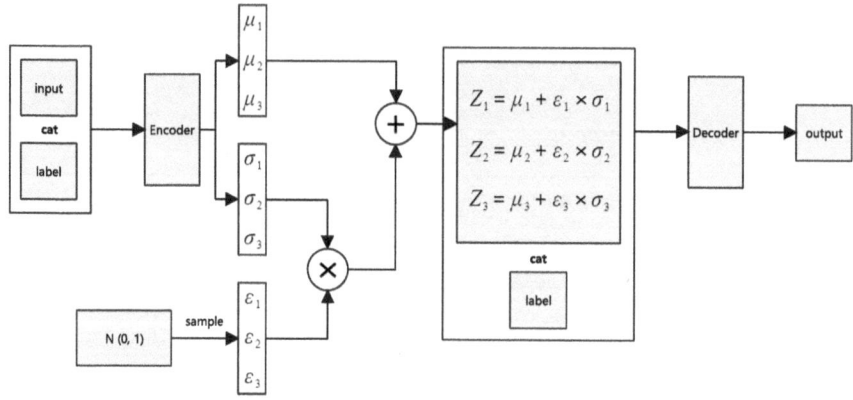

Fig. 4. The structure of CVAE

the decoder receives samples from the latent space and the same conditional information to reconstruct the original data. The training objective of CVAE is to maximize the Evidence Lower Bound (ELBO), which balances reconstruction accuracy with a regularization term that ensures the latent space follows a prior distribution (usually Gaussian). In this work, we use CVAE as the generator for a GAN to generate more realistic protocol messages.

3 Methodology

The purpose of this paper is to fuzz implementations of IoT protocols. An initial challenge is identifying efficient methods for generating test cases. In our work, we train a generative model on the dataset of protocol messages to generate spurious protocol messages that exhibit varying similarity with the authentic ones. The main steps of our approach are as follows.

3.1 Message Preprocessing

The real protocol messages that we capture need to be preprocessed before being used as training data for the model, as this data is not directly suitable for model training. We capture the real traffic between the server and the IoT devices using Wireshark [16]. The protocol messages are saved as a C file, which contains multiple arrays, each storing a hexadecimal message. Further processing and conversion are then needed to meet the input requirements of the model.

First, extract message sequences from the C file using regular expressions [8]. After extraction, remove unnecessary characters and fields, such as punctuation marks, comments, etc. Treat each message as a character sequence: $M_{1:N} = (m_1, m_2, ..., m_n, ..., m_N)$, $m_n \in S$, where S is the character set of messages. Algorithm 1 provides an overview of the process.

Algorithm 1 Extract message sequences

Require: T: C file
Ensure: $M_1, M_2, M_3, \ldots, M_n$: real message sequences
 1: $lines \Leftarrow$ RE('Array name', T)   ▷ Use a regular expression to find all matches.
 2: **for all** $line$ in $lines$ **do**
 3:    $line \Leftarrow$ RE('\\', $line$)   ▷ Remove comments.
 4:    $line \Leftarrow$ REMOVEMARKS($line$) ▷ Remove punctuation marks, '0x' and blank.
 5: **end for**
 6: $i \Leftarrow 1$ and Initialize $M_1, M_2, \ldots, M_n$ as empty strings
 7: **for all** $line$ in $lines$ **do**
 8:    **if** $line$ is not null **then**
 9:        $M_i \Leftarrow M_i + line$
10:        $i \Leftarrow i + 1$
11:    **end if**
12: **end for**
13: **return** $M_1, M_2, M_3, \ldots, M_n$

Second, stripping unnecessary message headers and alignment sequences. IoT protocols are typically built on top of other protocols like TCP or UDP [15]. Therefore, it is necessary to identify and remove the transport layer protocol (such as TCP header). Since the lengths of the protocol messages are not uniform, in order to standardize training for the model, we insert the special character "-" to fill the shorter sequences, which makes the length of all messages consistent with the longest length. Algorithm 2 provides an overview of the process.

Algorithm 2 Split and align real message sequences

Require: $M_1, M_2, M_3, \ldots, M_n$: real message sequences
Ensure: $A_1, A_2, A_3, \ldots, A_n$: aligned message sequences
 1: **for all** M_i in $\{M_1, M_2, M_3, \ldots, M_n\}$ **do**
 2:    $M_i \Leftarrow$ SPLIT(M_i)   ▷ Remove redundant headers.
 3: **end for**
 4: $l_{max} \Leftarrow$ FINDLENMAX($\{M_1, M_2, M_3, \ldots, M_n\}$)   ▷ Calculate the length of the
    longest message sequence.
 5: **for all** M_i in $\{M_1, M_2, M_3, \ldots, M_n\}$ **do**
 6:    $len \Leftarrow$ LENGTH(M_i)
 7:    **if** $len < l_{max}$ **then**
 8:        **for** $j \Leftarrow 0$ to $(l_{max} - len)$ **do**
 9:            $M_i \Leftarrow M_i +' -'$
10:        **end for**
11:    **else**
12:        $A_i \Leftarrow M_i$
13:    **end if**
14: **end for**
15: **return** $A_1, A_2, A_3, \ldots, A_n$

3.2   Message Clustering

After preprocessing the messages, we need to cluster them so that GCFuzz can learn the features of various categories of messages under supervision. This is because the lengths and specific field values of different message categories may vary. We introduce a clustering method based on a keyword for clustering protocol messages so that the clustered protocol messages have more similarities. So, finding the keyword is the work of this section. First, we divide each pair of characters in the aligned message sequence into a field , i.e. a byte. The result is shown in Fig. 5.

	K_1	K_2	K_3	K_4	K_5	K_6	K_7	K_8	K_9	K_{10}	K_{11}	K_{12}
m_1	14	7e	00	00	00	06	00	01	00	0f	00	04
m_2	0c	e3	00	00	00	06	00	02	00	3b	00	06
m_3	16	52	00	00	00	06	00	03	00	11	00	08
m_4	02	19	00	00	00	06	00	04	00	4e	00	08
m_5	08	0c	00	00	00	06	00	05	00	02	00	00
m_6	23	4f	00	00	00	06	00	06	00	57	a9	4c
m_7	25	b1	00	00	00	02	00	07	--	--	--	--
m_8	0d	a3	00	00	00	02	00	11	--	--	--	--

Fig. 5. Sequence division

Candidate Keywords. The most desired keyword to be selected should be the opcode in a message, as the opcode typically serves as the key identifier of the message, determining its type and operation. The selection of the opcode as the keyword makes it possible to recognize and understand the meaning of the message more efficiently.

For the reasons mentioned above, the keyword must be a dynamic field. Since the traffic we capture involves various operations, opcodes are not unique. Specifically, we need to exclude some fields that are equal in all messages. Fields containing "−" should also be excluded because the opcode is a field that should be present in all messages. Therefore, in Fig. 5, K_1, K_2, K_6, K_8 are considered as candidate keywords.

Final Keyword. Once candidate keywords have been selected, it is necessary to calculate the probability of these keywords being the final keyword. The keyword with the highest probability should be identified as the final keyword. The selection of the final keyword is primarily based on two key considerations:

– The message sequences within the same cluster should exhibit a high level of similarity.

– The messages within the same cluster should exhibit a high level of structural similarity.

1. **Similarity Score Between Messages in a Cluster**. We present Jaccard Similarity [20] for measuring the similarity between messages. The formula for calculating Jaccard Similarity can be expressed as the following formula 1:

$$\begin{cases} J_{ij} = \frac{|m_i \cap m_j|}{|m_i \cup m_j|} \\ P_1 = \frac{\sum_{i=1}^{n} \sum_{j=1}^{n} J_{ij}}{n^2} \end{cases} \quad (1)$$

where $|m_i \cap m_j|$ represents the number of elements in the intersection of sets A and B, and $|m_i \cup m_j|$ represents the number of elements in the union of sets A and B.

2. **Structure Similarity Score Between Messages in a Cluster**. We present the Edit Distance [24] for measuring the structural similarity between messages in a cluster. The formula for calculating Edit Distance can be expressed as the following formula 2:

$$\begin{cases} E_{ij} = 1 - \frac{\text{edit\_distance}(m_i, m_j)}{\text{max\_len}(m_i, m_j)} \\ P_2 = \frac{\sum_{i=1}^{n} \sum_{j=1}^{n} E_{ij}}{n^2} \end{cases} \quad (2)$$

where $edit\_distance(m_i, m_j)$ represents the minimum number of addition, deletion, or replacement operations required for m_i transform into m_j. Meanwhile, $max\_len(m_i, m_j)$ represents the length of the longest message sequence among m_i and m_j.

It is notable that if messages are clustered based on a specific keyword, and a category contains only one message, then the Message Similarity Score and the Structure Similarity Score Between Messages in a Cluster are considered to be 0. The purpose is to reduce the influence of message length on the choice of the keyword. Clustering based on message length may result in categories with a unique message. However, clustering based on the opcode would not lead to only one message in a certain category.

The probability of a candidate keyword becoming the final keyword is the average of the two metrics mentioned above. The candidate keyword with the highest score is the final keyword we want to choose. Once the final keyword has been determined, messages containing the same value of the keyword are clustered into a category. Finally, we get different categories of training data.

3.3   Determining the Model

In order to generate fake but credible test cases, we employ a GAN with a CVAE as the generative model, referred to as GC (GAN with CVAE). GC is a hybrid generative model that combines the principles of CVAE and GAN. In GC, the encoder and decoder of the CVAE are merged together as the generator of the GAN, which is used for learning the latent representations of the data and generating new data. Additionally, the discriminator of GAN helps to train the generator to generate more realistic data. The structure of GC is shown in Fig. 6. The specific components of GC are as follows:

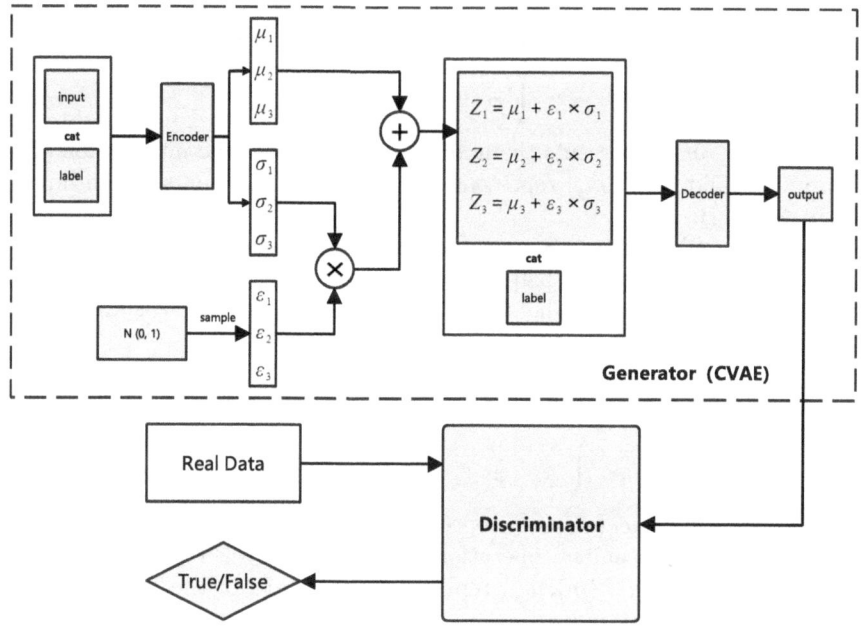

Fig. 6. The structure of GC

- **Encoder.** The encoder receives input data and conditional information through a fully connected layer [28], which maps them to a latent representation in the latent space. This representation is then processed by a ReLU activation function. Two additional fully connected layers are employed to obtain the mean μ and logarithm of the variance σ of the latent space.
- **Decoder.** The decoder receives sampled latent representations from the latent space and conditional information through a fully connected layer, which performs the decoding process, transforming the latent representations into reconstructed data. The activation function employed is ReLU. Subsequently, another fully connected layer is employed as the output layer, which generates data samples related to the conditional information. The activation function used in this layer is Sigmoid.
- **Discriminator.** The discriminator receives both real samples and generated samples (including conditional information) and attempts to distinguish

between them. It is composed of three fully connected layers, with ReLU used as the activation function for the first two layers and Sigmoid used as the activation function for the output layer.

– **Loss function.** The loss function of GC typically consists of two parts: the first is the reconstruction loss of CVAE, which ensures that the generated data is similar to the input data; the second is the adversarial loss of GAN, which encourages the generated data to match the distribution of real data. It is worth noting that during the training process, these two parts of the loss function are typically optimized simultaneously. On the one hand, during the update of the generator, the reconstruction loss of CVAE is employed to update the generator's parameters, with the objective of accurately reconstructing the input data. On the other hand, in the adversarial training of GAN, the generator and the discriminator utilize the adversarial loss to update their parameters respectively. The generator aims to generate more realistic samples, while the discriminator aims to improve its ability to distinguish between generated samples and real samples.

During the training phase, the GC is trained to generate samples that match the specified conditional information. This is achieved by inputting the relevant conditional information alongside the input data into the CVAE and the discriminator. The conditional information is labels for different categories of messages, which are transformed into one-hot encoding and spliced with the message sequence. Once the model is trained, new samples related to the given condition can be generated by generating a random noise and inputting it into the decoder together with the label.

4 Experiment

In this section, we evaluate our approach by testing a implementation of the Modbus-TCP protocol and comparing it with GANFuzz [11], Peach [7], BooFuzz [13], and Sulley [21].

4.1 Fuzzing Target

We evaluate GCFuzz by testing Modbus-TCP [29] protocol using MOD_RSSIM v8.20 (a popular implementation of the Modbus-TCP protocol) as the SUT. Modbus is an application layer protocol developed by Modicon in 1979. Nowadays, the Modbus protocol has become a standard communication protocol for many IoT environments. Therefore, testing it can assess the applicability and effectiveness of security tools and techniques in real IoT environments.

4.2 Evaluation Metrics

In order to evaluate the result of the experiment qualitatively and quantitatively, we employ the following four metrics.

- **Test case pass rate(TCPR).** TCPR means the percentage of test cases accepted by the SUT, regardless of whether these test cases conform to the protocol grammar. Note that these test cases may not receive a response, as they are merely recognized by the SUT as a Modbus-TCP request message. A higher TCPR indicates a better quality of test cases, indicating that the trained model has a rough grasp of the protocol grammar. The calculation formula of TCPR is as follows:

$$TCPR = \frac{Number\ of\ test\ cases\ accepted}{Total\ number\ of\ test\ cases} \tag{3}$$

- **Accuracy of Test case(AOTC).** AOTC means the accuracy of the generated test case. Protocol messages typically consist of multiple fields, and we judge the accuracy of test cases based on whether the message type or opcode field is correct. A higher AOTC indicates greater accuracy in the generated test cases. This metric can also be used to assess whether the model can accurately learn the protocol grammar. The calculation formula of AOTC is as follows:

$$AOTC = \frac{Number\ of\ test\ cases\ with\ correct\ opcode}{Total\ number\ of\ test\ cases} \tag{4}$$

- **Test case response rate(TCRR).** TCRR means the percentage of test cases that are accepted by the SUT and receive a response. If a test case is accepted by the SUT and receives a response, it is considered an accurate test case. A higher TCRR indicates better test case quality, indicating that the trained model has a good understanding of the protocol grammar, which helps to generate good fuzzing results. The calculation formula of TCRR is as follows:

$$TCRR = \frac{Number\ of\ test\ cases\ received\ a\ response}{Total\ number\ of\ test\ cases} \tag{5}$$

- **Accurate Fuzzing efficiency(AFE).** AFE means the number of test cases, which are accepted by the SUT and receive a response, inputted into the SUT per hour. A higher AFE indicates that a greater number of test cases conforming to the protocol grammar are generated per unit of time, resulting in greater accurate fuzzing efficiency. The calculation formula of AFE is as follows:

$$AFE = \frac{Number\ of\ test\ cases\ received\ a\ response}{Total\ time\ of\ fuzzing} \tag{6}$$

4.3   Training Dataset

The experimental training dataset was generated using the open source Modbus protocol library, Pymodbus [5]. Pymodbus is a library for the Python programming language, which is used for the Modbus communication protocol.

In this work, a Modbus-TCP protocol sending script was written using Pymodbus, which randomly sends 8 types of messages and receives data using MOD_RSSIM. Wireshark was used to monitor the communication process and record the communication traffic. Finally, we captured 1,000, 5,000, and 10,000 randomly generated request messages. It should be noted that generating messages in three different quantities was only for validating the feasibility of the clustering method based on a keyword. Ultimately, 10,000 messages were used to train the generative model.

4.4   Message Clustering

According to the method of clustering based on a keyword, we conducted experiments using different numbers of messages and calculated the probability of candidate keywords becoming the final keyword. The results of the experiments are shown in Table 1. According to the experimental results, the final keyword is K_8. And according to the actual Modbus-TCP message, the corresponding field of K_8 happens to be the opcode of the Modbus-TCP message. The message format of the Modbus-TCP protocol is shown in Fig. 7.

Table 1. Candidate keyword scores of Modbus-TCP message

	k_1	k_2	k_6	k_8
1000	62.26%	61.29%	69.76%	**72.10%**
5000	62.87%	65.14%	68.36%	**70.43%**
10000	63.94%	65.75%	68.38%	**70.18%**

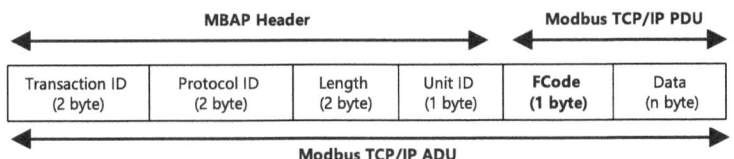

Fig. 7. The message format of Modbus-TCP

To further validate the feasibility of this clustering method, we also used MQTT [26] messages for verification. The results of the experiments are shown in Table 2. Based on the experimental results, the final keyword is K_1. And according to the actual MQTT message, the corresponding field of K_1 happens to be the opcode of the MQTT message, i.e., Message Type. The message format of the MQTT protocol is shown in Fig. 8. The feasibility of our clustering method has been confirmed based on the above experimental results.

Table 2. Candidate keyword scores of MQTT message.

	k_1	k_2
1000	**67.41%**	64.80%
5000	**79.42%**	73.63%
10000	**81.34%**	76.68%

bit	7	6	5	4	3	2	1	0
byte 1	Message Type				DUP	Qos Level		RET
byte 2	Remain Length							

Fig. 8. The message format of MQTT protocol

4.5   Model Training and Message Generation

After clustering, the messages were divided into 8 categories corresponding to 8 types of messages. We used a total of 10,000 messages from these 8 categories to train the model. Before training the model, it was necessary to balance the dataset, as the number of messages in each category is different due to the random generation of messages. To ensure fairness in model training, we used techniques such as oversampling and undersampling [19] to control the number of messages in each category to 1250.

We inputted these 10,000 messages and their corresponding labels into GC for training. The batch size was set to 16, the epochs were set to 20, and the model parameters and generated sequences were saved every 5 epochs. After training, we used the trained model to generate 10,000 messages, including 1250 messages for each of the 8 categories.

4.6   Experimental Results

In this section, we present the bugs and exceptions found during testing and evaluate the results using the TCPR, AOTC, TCRR, and AFE metrics. Meanwhile, we make a comparison of four training stages. The best trained model is used to compare with GANFuzz, Peach, BooFuzz, and Sulley.

Comparing Models with Different Training Epochs. We compared the fuzzing results of GCFuzz under different epochs. Due to the randomness of the model during message generation, which affects the final fuzzing results, we repeated each item 10 times and took the average as the final experimental result [14]. Figure 9 illustrates the values of four metrics during the alternating training process from 5 to 20 epochs. GCFuzz, with the best training results at 20 epochs, will be compared with other fuzzers.

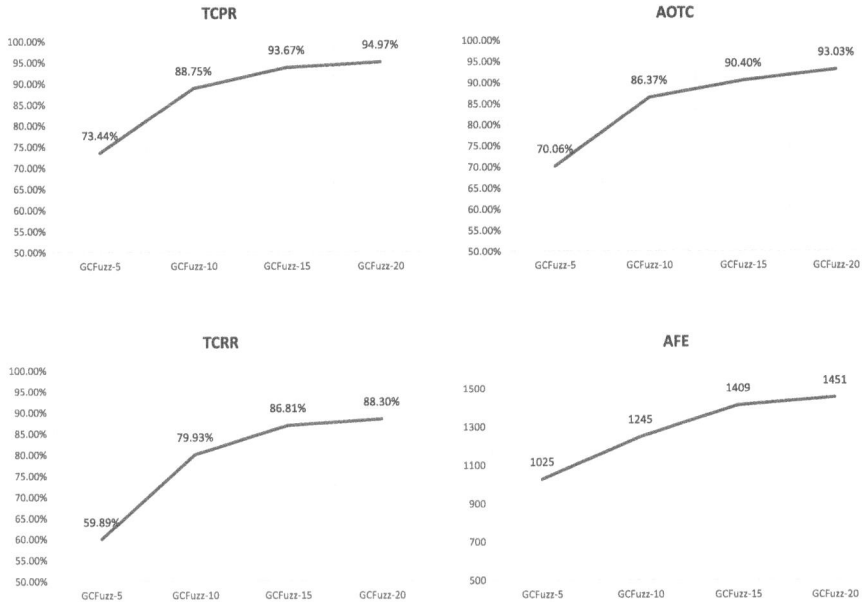

Fig. 9. TCPR, AOTC, TCRR, and AFE values in different epochs

Comparing Models with Other Fuzzers. We recorded the experimental time of the model trained with 20 epochs, including model training, message generation, and message transmission, totaling 6 h. We then performed 6 h of fuzzing Modbus-TCP using GANFuzz, Peach, BooFuzz, and Sulley respectively. Similarly, we ran each item 10 times and took the average as the final experimental result. The experimental results are shown in Fig. 10.

The experimental results show that the TCPR of GCFuzz is not as high as 100% like BooFuzz and Sulley, but its AOTC is much higher than Peach, BooFuzz, and Sulley. AOTC refers to the accuracy of the generated test cases, which means that the test cases generated by GCFuzz are better than traditional generation-based protocol fuzzers. This is because GCFuzz does not generate messages randomly but instead generates fake but credible messages based on real protocol messages. In addition, TCRR of GCFuzz is much higher than these traditional generation-based protocol fuzzers and GANFuzz too. TCRR refers to the percentage of test cases that are accepted by the SUT and receive a response, which means that the test cases generated by GCFuzz are largely compliant with the protocol grammar, saving a significant amount of computational resources that would otherwise be wasted on invalid test cases. Since GCFuzz requires no prior knowledge and can generate a large number of test cases based on real network traffic, it not only saves a lot of manpower and time, but also makes the testing process more universal and suitable for other protocols. Furthermore, the AFE of GCFuzz is also much higher than Peach, BooFuzz, and Sulley, which means that GCFuzz has greater accurate fuzzing efficiency than these fuzzers.

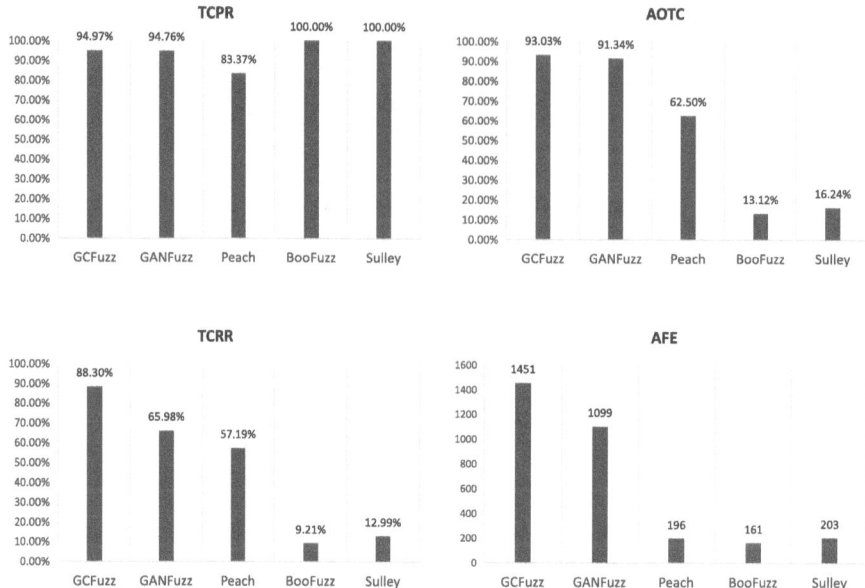

Fig. 10. TCPR, AOTC, TCRR, and AFE values in different fuzzers

And we have not taken into account the time required to provide prior knowledge for these fuzzers (i.e. writing the appropriate configuration files). Considering this point, our approach obviously has a significant advantage.

In summary, compared to GANFuzz, Peach, BooFuzz, and Sulley, the test cases generated by our method have stronger targeting and less redundancy, while our method also has higher accurate testing efficiency.

4.7   Bug Analysis

This section analyzes the bugs we found. During fuzzing, we did not observe any abnormal program terminations. However, it is worth noting that some of the test cases generated by GCFuzz had message lengths that did not match their length fields, but MOD_RSSIM considered them valid and responded normally. This is a typical bug caused by inconsistencies in protocol grammar. It is also the reason why MOD_RSSIM often replies with a "timeout" exception. When MOD_RSSIM receives messages from GCFuzz, it does not receive the message directly based on the length field in the message. Instead, it stores the whole message in a data buffer and then receives the contents of the message in segments. Specifically, MOD_RSSIM first reads the header field of the message from the buffer, then determines the length of its data field based on the content of the header field, and finally reads the data from the data buffer. However, because the extracted data does not match the actual data length, this will cause MOD_RSSIM to enter a lock state, resulting in a "timeout" exception.

In addition, we have found that MOD_RSSIM has detection errors regarding protocol opcodes when receiving requests. For example, opcode "01" in the protocol message indicates a read coil operation. GCFuzz generated a protocol message with an opcode of "81", which MOD_RSSIM incorrectly interpreted as a read coil operation. This could be due to programming logic errors within the MOD_RSSIM software or a lack of understanding of the protocol grammar. Alternatively, it could be due to Wireshark's poor understanding of the Modbus-TCP protocol leading to errors in traffic parsing. Fortunately, although MOD_RSSIM receives the request, it does not provide a corresponding response but rather replies that this is an illegal function.

5   Related Work

Generation-Based Fuzzers. For generation-based fuzzers such as Peach [7], BooFuzz [13], and Sulley [21], test cases are generated through a manually constructed test model. The test model describes the syntactic structure of the protocol, such as field types, field sizes, and valid value ranges. Users need to provide these models by analyzing source code or reading protocol specifications. SPFuzz [27] has made further improvements by providing specific mutation strategies for different fields to obtain higher quality test cases. However, constructing these specifications can be a laborious and manpower-intensive task, and the manually defined test model may not necessarily generate high-quality test cases well. In contrast, we utilize deep learning to learn the message structure of the protocol. By using real network traffic, GCFuzz can generate a large number of high-quality test cases with its powerful learning capabilities.

Mutation-Based Fuzzers. Mutation-based fuzzers, such as AFLNet [22], generate test cases by selecting an existing input from the seed pool and performing random mutations. Since there is no need to know the protocol specification and message format in advance, these fuzzers are easy to implement. However, due to the lack of format specifications, these fuzzers may quickly encounter obstacles, because the packets generated by blind mutations may be discarded by highly structured protocol implementations.

Intelligent Fuzzers. With the development of artificial intelligence(AI), researchers have begun to apply AI technologies to the field of fuzzing. For example, Learn&fuzz [9] uses the seq2seq model to learn the grammar of PDF (a complex file format) objects and uses the trained model to generate test cases for testing PDF parsers. Rajpal et al. [23] employed a neural network model to predict optimal locations within input files for fuzzing mutations. GANFuzz [11] employs GAN [10] to grasp protocol grammar based on real protocol messages and generate fake but credible messages using the trained generative model. These efforts have made a number of contributions to network security. Different from the existing work, we propose an intelligent approach for generating test cases for IoT protocols by employing a supervised learning model. Specifically, we use a CVAE as the generator of a GAN, so that it can learn labeled data.

6   Conclusion and Future Work

This paper presents an intelligent test case generation method for fuzzing IoT protocols. First, to improve message clustering, we propose a clustering method based on a keyword. The effectiveness of our method is verified by experiments using messages from Modbus-TCP and MQTT protocols. Second, to better capture the features of messages from different categories, we use a GAN with CVAE, where a CVAE serves as the generator for the GAN. Experimental results show that GCFuzz can generate high-quality test cases without any prior knowledge. Compared to GANFuzz, Peach, BooFuzz, and Sulley, GCFuzz generates test cases with stronger targeting and less redundancy, while GCFuzz also has higher testing efficiency.

In the future, our work will focus on the following directions. Firstly, we aim to apply our method to stateful protocols such as FTP, RTSP, and others. Secondly, we plan to integrate grey-box fuzz testing techniques to improve the accuracy of testing. Lastly, we intend to leverage artificial intelligence technology for automating the analysis of discovered vulnerabilities.

Acknowledgments. This work was partly supported by the National Natural Science Foundation of China under Grant nos.62072412 and 61902359, the Opening Project of Shanghai Key Laboratory of Integrated Administration Technologies for Information Security under Grant AGK2018001, the National Key Research and Development Program of China under Grant no.2019YFC0118800, the Zhejiang Province Science Foundation under Grants LD24F020002, LQ24F020025, and the Jinhua Science and Technology Plan under Grant 2023-1-091.

References

1. Al-Sarawi, S., Anbar, M., Alieyan, K., Alzubaidi, M.: In: 2017 8th International conference on information technology (ICIT), pp. 685–690. IEEE (2017)
2. Asha, P., et al.: IoT enabled environmental toxicology for air pollution monitoring using AI techniques. Environ. Res. **205**, 112574 (2022)
3. Baezner, M., Robin, P.: Stuxnet. Technical report, ETH Zurich (2017)
4. Bangemann, T., et al.: State of the art in industrial automation. Industrial Cloud-Based Cyber-Physical Systems: The IMC-AESOP Approach, pp. 23–47 (2014)
5. Collins, G.: Pymodbus documentation (2013)
6. Creswell, A., White, T., Dumoulin, V., Arulkumaran, K., Sengupta, B., Bharath, A.A.: Generative adversarial networks: an overview. IEEE Signal Process. Mag. **35**(1), 53–65 (2018)
7. Eddington, M.: Peach fuzzing platform. https://peachtech.gitlab.io/peach-fuzzer-community. Accessed 08 Nov 2024
8. Friedl, J.E.: Mastering Regular Expressions. O'Reilly Media, Inc., Sebastopol, California (2006)
9. Godefroid, P., Peleg, H., Singh, R.: Learn&fuzz: machine learning for input fuzzing. In: 2017 32nd IEEE/ACM International Conference on Automated Software Engineering (ASE), pp. 50–59. IEEE (2017)
10. Goodfellow, I., et al.: Generative adversarial nets. In: Advances in neural Information Processing Systems, vol. 27 (2014)

11. Hu, Z., Shi, J., Huang, Y., Xiong, J., Bu, X.: GANfuzz: a GAN-based industrial network protocol fuzzing framework. In: Proceedings of the 15th ACM International Conference on Computing Frontiers. pp. 138–145 (2018)

12. Jansen, D., Buttner, H.: Real-time ethernet: the ethercat solution. Comput. Control. Eng. **15**(1), 16–21 (2004)

13. Jtpereyda: Boofuzz: network protocol fuzzing for humans. https://github.com/jtpereyda/BooFuzz. Accessed 08 Nov 2024

14. Klees, G., Ruef, A., Cooper, B., Wei, S., Hicks, M.: Evaluating fuzz testing. In: Proceedings of the 2018 ACM SIGSAC Conference on Computer and Communications Security, pp. 2123–2138 (2018)

15. Kumar, S., Rai, S.: Survey on transport layer protocols: TCP & UDP. Int. J. Comput. Appl. **46**(7), 20–25 (2012)

16. Lamping, U., Warnicke, E.: Wireshark user's guide. Interface **4**(6), 1 (2004)

17. Miller, B.P., Fredriksen, L., So, B.: An empirical study of the reliability of Unix utilities. Commun. ACM **33**(12), 32–44 (1990)

18. Miller, B.P., et al.: Fuzz revisited: a re-examination of the reliability of Unix utilities and services. Technical report, University of Wisconsin-Madison Department of Computer Sciences (1995)

19. Mohammed, R., Rawashdeh, J., Abdullah, M.: Machine learning with oversampling and undersampling techniques: overview study and experimental results. In: 2020 11th International Conference on Information and Communication Systems (ICICS), pp. 243–248. IEEE (2020)

20. Niwattanakul, S., Singthongchai, J., Naenudorn, E., Wanapu, S.: Using of Jaccard coefficient for keywords similarity. In: Proceedings of the International Multiconference of Engineers and Computer Scientists, vol. 1, pp. 380–384 (2013)

21. OpenRCE: Sulley. https://github.com/OpenRCE/sulley. Accessed 08 Nov 2024

22. Pham, V.T., Böhme, M., Roychoudhury, A.: Aflnet: a GreyBox fuzzer for network protocols. In: 2020 IEEE 13th International Conference on Software Testing, Validation and Verification (ICST), pp. 460–465. IEEE (2020)

23. Rajpal, M., Blum, W., Singh, R.: Not all bytes are equal: neural byte sieve for fuzzing. arXiv preprint arXiv:1711.04596 (2017)

24. Ristad, E.S., Yianilos, P.N.: Learning string-edit distance. IEEE Trans. Pattern Anal. Mach. Intell. **20**(5), 522–532 (1998)

25. Rose, K., Eldridge, S., Chapin, L.: The internet of things: an overview. Internet Soc. (ISOC) **80**(15), 1–53 (2015)

26. Singh, M., Rajan, M., Shivraj, V., Balamuralidhar, P.: Secure MQTT for internet of things (IoT). In: 2015 Fifth International Conference on Communication Systems and Network Technologies, pp. 746–751. IEEE (2015)

27. Song, C., Yu, B., Zhou, X., Yang, Q.: SPfuzz: a hierarchical scheduling framework for stateful network protocol fuzzing. IEEE Access **7**, 18490–18499 (2019)

28. Sun, D., Wulff, J., Sudderth, E.B., Pfister, H., Black, M.J.: A fully-connected layered model of foreground and background flow. In: Proceedings of the IEEE Conference on Computer Vision and Pattern Recognition, pp. 2451–2458 (2013)

29. Swales, A., et al.: Open MODBUS/TCP specification. Schneider Electric **29**(3), 19 (1999)

30. Tech, P.: Peach fuzzer configuration file (peach pit). https://peachtech.gitlab.io/peach-fuzzer-community/v3/PeachPit.html. Accessed 08 Nov 2024

31. Zhang, C., Barbano, R., Jin, B.: Conditional variational autoencoder for learned image reconstruction. Computation **9**(11), 114 (2021)

32. Zhong, R.Y., Xu, X., Klotz, E., Newman, S.T.: Intelligent manufacturing in the context of industry 4.0: a review. Engineering **3**(5), 616–630 (2017)

VRMDA: Verifiable and Robust Multi-subset Data Aggregation Scheme in IoT

Jianying Li[1], Shuo Zhou[2(✉)], and Yining Liu[2]

[1] School of Computer and Information Security, Guilin University of Electronic Technology, Guilin 541004, China
[2] School of Data Science and Artificial Intelligence, Wenzhou University of Technology, Wenzhou, Zhejiang 325035, China
20035056@qq.com

Abstract. Data aggregation masks individual entries by summing values, thus preventing the disclosure of private information. Multi-subset data aggregation divides subsets according to the range of data and collects the aggregated values of each subset. This provides the data control center with granular data. If data aggregation loses its verifiability, it becomes impossible to confirm the authenticity of messages or detect any tampering. This vulnerability could lead to an increased incidence of data tampering attacks across the system. Furthermore, a centralized control center is prone to disruptions due to natural disasters or malicious attacks, creating a single point of failure that may cause service interruptions and data loss. This could result in service interruptions and data loss. Therefore, we propose a robust and verifiable multi-subset data aggregation scheme. In this scheme, ciphertext information is transmitted to multiple control centers in the form of shards, and the aggregation results are recovered through cooperation between control centers. The security analysis demonstrates the protocol could resist collusion attacks and single points of failure. Experimental analysis indicates that the scheme is feasible.

Keywords: Data Aggregation · Data security · Smart grid · Verifiable

1 Introduction

In today's rapidly developing digital age, the smart grid, as the future form of the power system [5], carries the important mission of energy transformation and optimizing resource allocation [8]. Smart meters, continuously generate a large amount of electricity consumption data. The collection and analysis of electricity data [20] are of great value for predicting the power generation of the entire power system [6], optimizing resource allocation, energy transformation, and formulating electricity prices [2]. However, the collection of real-time data may potentially expose users' lifestyle habits [10]. To address this issue, researchers have proposed various data aggregation and privacy-preserving schemes.

W. Meng et al. (Eds.): ADIoT 2024, LNCS 15397, pp. 126–138, 2025.
https://doi.org/10.1007/978-3-031-85593-1_8

The advantage of privacy-preserving data aggregation lies in its ability to provide data collectors with comprehensive statistical information, such as totals or averages [18]. [1,7,23] studies improved fine-grained aggregation. In existing research schemes, to balance data and privacy, most researchers use techniques such as homomorphic encryption, multi-party secure computation, and increased noise.

Verifiability is the foundation for ensuring the accuracy and correctness of data aggregation results. The security issues in the smart grid are noteworthy [14]. External adversaries may eavesdrop and intercept the data transmitted between two logical entities in a data aggregation system, and further disrupt or replace critical information, thereby compromising the integrity of communication messages. Verifiable data aggregation has been considered a promising solution.

What's more, if the data control center encounters network attacks or hardware failures, it may cause power supply interruptions, affect the stability and reliability of the power grid, and even trigger large scale power outages, causing significant impacts on the social economy and residents' lives. Establishing multiple data control centers can significantly improve the security and reliability of data management.

Therefore, our research will focus on verifiability and avoidance of single point of failure, the main contributions are as follows:

- Verifiable. This article employs Shamir's Secret Sharing and verifiable technology to ensure the verifiability and security of data. It enhances the system's security and trustworthiness and rectifies potential security vulnerabilities.
- Robust. The architecture incorporates multiple control centers, which adapt to their dynamic fluctuations, thereby bolstering the security of data management and mitigating the risk of single points of failure. This design also efficiently thwarts potential collusion from internal sources, safeguarding the system's stability and the integrity of the data.
- Privacy-preserving. Employing techniques such as Paillier homomorphic encryption, it is possible to manipulate encrypted data without the need for decryption. This approach facilitates fine-grained data aggregation, striking a balance between preserving data privacy and ensuring its utility.

The rest of this paper is organized as follows: Sect. 2 discusses the related work. In Sect. 3, we introduce our system model including the communication model, threats model, and design objective. We present our proposed scheme in Sect. 4, followed by security and performance analysis in Sect. 5. Sect. 6 provides a conclusion.

2   Related Work

In recent years, a great deal of research has been conducted around data aggregation, with the verifiability of aggregation being crucial security issues in data aggregation.

In 2021, Yan et al. [21] emphasized that the reliability of fog nodes is crucial for the accuracy of data aggregation results and that data transmitted by these nodes must undergo verification before being accepted by the server. Their proposed scheme enables requesters to verify the correctness of aggregation results and tolerates a few faulty Fog Nodes (FNs) without compromising the data aggregation outcome. Further enhancing security, Manikandan et al. [12] introduced a multi-parameter secure data aggregation technique tailored for data center environments. This scheme integrates integrity verification technology to perform robust aggregation by securely combining vector elements within the data center. In 2023, Zhang et al. [25] presented the LAVODA scheme, which ensures data confidentiality while providing verifiability of the aggregation process. Specifically, the scheme utilizes dual Diffie-Hellman (DH) in conjunction with circle-based location verification to safeguard the privacy of the requester's location policy and the data provider's location. As described in [19], the scheme focuses on achieving efficient maritime transportation by employing zero-knowledge proof technology to perform continuous attribute authentication of maritime terminals. Similarly, Chen et al. [3] developed a fine-grained linear homomorphic encryption scheme capable of assigning varying weights to different dimensions of user data while preserving linear homomorphic properties across each dimension. Regarding 2024, Zhao et al. [26] designed the VMEMDA scheme, which facilitates data aggregation that can selectively process data types, such as spatial or temporal data aggregation. This scheme aggregates multi-source encrypted medical data into a single ciphertext.

Single points of failure pose significant risks to both system reliability and business continuity. The presence of a sole control center within a system greatly increases its vulnerability to such failures.

As early as 2021, Yang et al. [22] addressed the concern of single points of failure within fog nodes. However, the current discourse seldom acknowledges the single point of failure risk in control centers. Given that extreme weather events and potential attacks can compromise the control center and alter the smart grid's structure, enhancing the reliability of data aggregation schemes is of paramount importance.

In 2021, Merad-Boudia et al. [13] introduced an efficient and secure multi-dimensional aggregation scheme named ESMA. Within ESMA, multi-dimensional data is structured and encrypted into a single Paillier ciphertext, streamlining the aggregation process. Peng et al. [16], in 2022, presented a scheme that utilizes a Chinese Remainder Theorem transformation method coupled with a counter to encode multi-dimensional data into a large integer, suitable for operations by a linear homomorphic encryption scheme. Also in 2022, Peng et al. [17] proposed a Privacy-Preserving Multi-Dimensional Aggregation (PMDA) scheme for multi-dimensional data. This scheme employs batch verification methods to ensure device data's non-repudiation and enhance edge node verification efficiency. In 2023, Liu et al. [9] introduced a multi-dimensional data aggregation scheme named OPERA, designed to boost users' willingness to share data. Concurrently, Chen et al. [4] proposed a resilient data aggregation scheme for smart grids, capable of adapting to dynamic structural changes within the smart grid. To counteract false data injection attacks, Pang et al. [15] constructed an efficient and privacy-preserving scheme that supports secure smart grid operations and safeguards data aggregation communication against malicious activities.

In 2024, Ma et al. [11] introduced a pioneering Dynamic Robust Iterative Filtering (DRIF) mechanism designed to enhance IoT application service quality and strengthen data aggregation resilience against collusion attacks.

However, the aforementioned works fail to address both the public verifiability of transmitted data and the risk of a single point of failure at the control center. In contrast, this scheme addresses these two critical issues, enabling the achievement of fine-grained data aggregation with enhanced reliability.

3   System Model

In this section, we delve into a comprehensive description of the system model, encompassing the communication framework, threat paradigm, protocol design objectives, and the symbolic notation used for representation (Table 1).

3.1   Communication Model

The system comprises four key components: smart meters at the user's premises, data gateways, central control centers, and an offline trusted authority.

Control Center (CC): Assuming the number of CC is 's', they are all responsible for receiving data shards from the aggregator. Everyone is semitrusted and honestly executes the protocol, but attempts to find and crack privacy information from legitimate data.

Gateway (GW): GW is a semitrusted workstation that serves as an intermediate node to perform three functions: data aggregation, data sharing, and data transmission.

Smart Meters (SM): The number of SMs the system assumes is 'n'. These smart meters can record real-time power consumption data and are responsible for encrypting the collected power data according to the requirements of the control center and uploading it to the GW.

Trusted Agency (TA): TA does not participate in the aggregation process. It is responsible for publishing parameters through a secure channel. After allocating them to the user end and control center end, cancel the network connection.

3.2   Threats Model

This scheme adopts a semihonest security model, which means that entities within the system will strictly and firmly enforce the agreement but attempt to infer private information from the information they receive. This scheme mainly considers collusion attacks and single points of failure to protect the privacy of user data and ensure data integrity.

Collusion Attacks: Entities within the system will strictly and firmly enforce the agreement but attempt to infer private information from the information they legally receive.

1. Malicious SMs will attempt to collude with GW and a certain CC_i in order to obtain personal data of other SMs.
2. CC_i infers that individual user data will attempt to collude with GW.
3. CC_i attempts to collude with each other to recover the decryption key.

Single Point of Failure: The control centers are susceptible to single-point failures. We will consider one scenario where such failures occur.

4. Control centers are susceptible to single points of failure due to natural disasters. They may lose the ability to send correct data reports due to natural disasters, power supply failures, or malicious attacks by adversaries.

3.3   Design Goals

The scheme is designed to implement a multi-subset aggregation protocol that resists collusion attacks, ensuring the following properties: integrity, verifiability, and privacy preservation.

1. Integrity: During the operation of the protocol, a large amount of data is exchanged through open channels. Integrity aims to ensure that data is complete and valid throughout the entire transmission process.
2. Verifiability: The system has designed a zero-knowledge proof verification process based on Shamir secret sharing technology to ensure that the ciphertext shards collected by the control center are true and accurate.
3. Privacy-persevering: Within the system, apart from the users themselves, even if any external or internal attacker intercepts the encrypted data mentioned above, they cannot obtain privacy information.

3.4   Symbol Description

Table 1. Symbol description

Sign	Symbol description
CC	Control Center
GW	Gateway
TA	Trusted Agency
SM	Smart meter
U_i	the i-th user
c_i	the i-th user's ciphertext
$|U_j|$	Number of users in the j-th subset
$[R_j, R_{j+1})$	the j-th subset
t	Number of Control Centers
k	Number of subsets
$\mathbb{U}$	Collection of n users

4   Data Aggregation Scheme

In this section, there are 7 stages that the system needs to execute, we will provide a detailed introduction to the entire process of the protocol through 7 stages.

4.1   System Initialization

The scheme goes through 5 steps during the system initialization process

Step1: TA chooses two large prime numbers p and q (length is 512 bit). Then TA produces public key $(N = pq, g)$ and private key (λ, μ) in *paillier* system. Here, $\lambda = lcm(p - 1, q - 1)$. The private key is divided into s shares using Shamir's secret sharing method, and only $t(1 < t <= s)$ of those shares are required to recover the complete private key.

Step2: In order to prevent collusion, TA generates a series of pseudo-random numbers $\{x_0, x_1, x_2, \cdots, x_n\}$ from the domain $\mathbb{Z}_N^*$, with pseudo-random number generator. These numbers meet condition $\sum_{i=0}^{n} x_i = 0 \bmod \lambda (i = 1, 2, 3, \cdots, n)$.

Step3: Through secure channels, TA partitions x_0 using the same Shamir method and assigns them to CCs terminals respectively, sends x_i to U_i and publishes public key (N, g).

Step4: TA generate a set of super-increasing sequence numbers $\{a_1, a_2, \cdots, a_k\}$, which satisfied $a_j > \sum_{j=1}^{i-1} a_j(R_{j+1} - R_j)n$, $(j = 2, 3, \cdots, k)$, where $a_1 = 1$. Then TA uses k intervals $[R_1, R_2), [R_2, R_3), \cdots, [R_k, E)$, divided according to fine-grained requirements, to generate $\{g^{a_1}, g^{a_2}, \cdots, g^{a_k}\}$ and publish these information.

Step5: TA takes the large prime number p, and the holders of sub keys are each CC, with a quantity of s. $'t'$ is the threshold value. $\mathbb{Z}_p^*$ is a finite field of order p. TA randomly selects $t - 1$ numbers from the finite field $\mathbb{Z}_p^*$, denoted as $\{b_1, b_2, \cdots, b_{t-1}\}$, As a coefficient of a $t - 1$ degree polynomial non polynomial term. The secrets are $\mathbf{X} = \{x_0, \lambda, \mu\}$, respectively. Polynomials are $\mathbf{F(x)} = \mathbf{X} + \sum_{l=1}^{t-1} b_l x^l (l = 1, 2, 3, \cdots, s)$. These s holders are respectively referred to as $CC_1, CC_2, \cdots, CC_s$, the sub keys assigned to each CC are $\mathbf{F(l)} = \{f(l), g(l), \text{and } h(l)\}$.

4.2   Data Encryption

Set $\mathbb{U} = \{U_1, U_2, U_3, \cdots, U_n\}$ represents n users. For every U_i, his electricity consumption data m_i meets the condition $m_i \in [R_i, R_{i+1})$. The ciphertext generated by the SM is

$$c_i = g^{a_j} \cdot H(t)^{N \cdot x_i} \bmod N^2. \tag{1}$$

Besides, the scheme define data increment as $\Delta m_i = m_i - R_i + x_i$, $m_i \in [R_i, R_{i+1})$. In the communication between the users and the GW, every user encrypts the parameter a_j representing the j-th subset and transmits it to the GW with the data increment of his power data. The transmission information is $\{c_i \| \Delta m_i\}$.

4.3 Data Aggregation

After receiving the message, GW aggregates ciphertext as follows

$$C = \Pi c_i = g^{a_1|U_1|+a_2|U_2|+\cdots+a_k|U_k|} \cdot H(t)^{N \cdot \sum_{i=1}^{n} x_i} \bmod N^2. \tag{2}$$

Besides, by aggregating data increments, SUM' can be obtained.

$$SUM' = \sum_{i=1}^{n} \Delta m_i = \sum_{i=1}^{n}(m_i - R_i) + \sum_{i=1}^{n} x_i. \tag{3}$$

GW performs data fragmentation on the aggregated ciphertext $C = \Pi c_i$ and aggregated data increment SUM' according to TA, and transmits the segmented s-slice data to CCs separately.

4.4 Data Ciphertext Fragmentation

TA sends the following information to GW before disconnection: the number of CC terminals is s, and the threshold value is t. $\mathbb{Z}_p^*$ is a finite field of order p. GW randomly selects $t-1$ numbers from the finite field $\mathbb{Z}_p^*$, denoted as $\{d_1, d_2, \cdots, d_{t-1}\}$, As a coefficient of a $t-1$ degree polynomial non polynomial term. The secrets are $C = \Pi c_i$ and SUM'. Polynomials are $C(x) = C + \sum_{l=1}^{t-1} d_l x^l$ and $Sum(x) = SUM' + \sum_{l=1}^{t-1} d_l x^l$. The sub keys assigned to these holders are $C(l)$ and $Sum(l)$.

4.5 Publicly Verifiable

Taking the ciphertext $C(l)$ after shard aggregation as an example, GW performs the following process. GW generates two large prime numbers p and q satisfying $p = 2q + 1$, define G_q as a cyclic subgroup Z_q of order q of the group, find the generator g_1 and g_2. GW selects a random number r in Z_q^*, registers $y = g_1^r$ as his public key, and chooses a hash function $hash : \{0,1\} \rightarrow Z_q^*$.

GW calculates ciphertext $C(l)$ for sub share and publishes sub share $Y = y^{C(l)}$. For all coefficients $\{d_1, d_2, \cdots, d_{t-1}\}$ of polynomial $C(x)$, declare commitment $Covenant = g_2^C, g_2^{d_1}, g_2^{d_2}, \cdots, g_2^{d_{t-1}}$.

The prover GW needs to prove that the sub share ciphertext Y is the correct encryption of the sub share $C(l)$. Firstly, calculate $X = \prod_{l=1}^{t-1} g_2^{d_l} \cdot g^C$ using publicly available information. Next, GW needs to prove that packet $\{g_2, X, y, Y\}$ satisfy the condition $X = g_2^{C(l)}$, $Y = y^{C(l)}$ and then GW performs the following operations:

Step 1: GW choose a random number $r \in Z_q^*$, building commiments $w_1 = g_2^r$, $w_2 = y^r$.
Step 2: GW calculates the challenge value $u = hash(w_1, X, w_2, Y)$.

Step 3: GW calculates the response value $v = r + f(l)u \bmod q$.

Step 4: The complete zero-knowledge proof $proof^2(w_1, w_2, u, v)$ obtained by GW will be published, and Anyone can verify whether the equation holds for each participant based on this proof and system parameters, $g_2^v = w_1 \cdot X^u$ and $y_2^v = w_2 \cdot Y^u$.

If it is true, it indicates that the participant performed the correct operation (the aggregator performed the correct Shamir sharing operation and shared the correct data shards). Otherwise, the participant is considered dishonest and removed from this protocol.

4.6 Data Shard Recovery

Each CC utilizes the homomorphic properties of secret sharing to extract all sub shares ciphertexts Y_C sent to it from the public information and saves them for subsequent recovery of ciphertexts. only $t(1 < t <= s)$ of those CCs are required to recover the complete private key. Use the following equation,

$$f(x) = \sum_{i=1}^{t} f(i) \cdot \prod_{j=1, j \neq i}^{t} \frac{x - x_j}{x_i - x_j}. \tag{4}$$

By substituting $x = 0$, the aggregated ciphertext C can be recovered, and similarly, the aggregated data increment SUM' can be recovered; Restore homomorphic decryption key (λ, μ) and restore blind factor x_0.

4.7 Data Decryption

Thus, the ciphertext decryption process is carried out by substituting the blind factor x_0 into the following equation

$$V = C \cdot H(t)^{N \cdot x_0} \tag{5}$$

$$= g^{a_1|U_1| + a_2|U_2| + \cdots + a_k|U_k|} \cdot H(t)^{N \cdot \sum_{i=0}^{n} x_i} \bmod N^2$$

$$= g^{a_1|U_1| + a_2|U_2| + \cdots + a_k|U_k|} \bmod N^2.$$

As represented by $D = a_1|U_1| + a_2|U_2| + \cdots + a_k|U_k|$, it can be obtained that

$$V = C \cdot H(t)^{N \cdot x_0} = g^D \bmod N^2.$$

CC can calculate D through the equation

$$D = \frac{V - 1}{N} \bmod N^2. \tag{6}$$

After obtaining the decrypted D, the CC needs to use Algorithm 1 to obtain the data distribution.

Then, CCs calculate as follows:

Algorithm 1 Post-decryption

Input: D, $a_j(j = 1, 2, \cdots, k)$
Output: $|U_j|$
  1: **for** $j = k$ to 1 **do**
  2:    $|U_j| = (D - D \bmod a_j)/a_j$
  3:    $D = D - (a_j|U_j|)$
  4: **end for**
  5: **Return** $|U_1|, |U_2|, \cdots, |U_k|$

$$SUM^* = R_1|U_1| + R_2|U_2| + \cdots + R_k|U_k|. \tag{7}$$

To obtain the sum of data increments, the aggregate value SUM' is processed as follows:

$$SUM'' = (SUM' + x_0) \bmod \lambda = \sum_{i=1}^{n}(m_i - R_i). \tag{8}$$

At the end of this stage, CC can calculate the total electricity consumption:

$$SUM = SUM^* + SUM''. \tag{9}$$

5 System Analysis

In this section, we will analyze the security of the system and elaborate on the feasibility of the system experiment.

5.1 Security Analysis

In this section, we conduct a security analysis of the protocol. The designed scheme meets the previously stated design objectives and can withstand various threats outlined in the threat model.

Theorem 1: Having resistance to collusion attacks means that after data is legally transmitted, honest and curious entities cannot obtain users' basic data information through internal collusion.

Scenario 1: SM_i collude with CC_i or GW in an attempt to obtain data from individual users. And this is not feasible.

Proof 1: After system initialization, SM_i has data includes x_i, public key (N, g), ciphertext c_i. CC_i has Shamir's data sharding $Share_{c_i}$ and $Share_{\Delta m_i}$. When CC_i conspires with a certain SM_i, CC_i can only obtain SM_i's personal data. GW can obtain c_i, Δm_i, $C = \Pi c_i$, Sum. When GW conspires with a certain SM_i, it cannot decrypt a single ciphertext due to the lack of a complete decryption key and blind factor x_i. Therefore, it is impossible to obtain confidential data from other users. This meets the design goals of privacy protection and integrity that we previously proposed.

Scenario 2: The GW and CC_i collude to be unable to obtain individual user data.

Proof 2: GW can obtion c_i, Δm_i, $C = \Pi c_i$, Sum. CC_i has Shamir's data sharding $Share_{c_i}$ and $Share_{\Delta m_i}$. CC_i cannot obtain the complete decryption key and blind factor x_i, so it cannot obtain the secret data of individual users.

Scenario 3: Collusion between several control centers attempting to obtain personal user data is not feasible. The setting of multiple control centers can effectively withstand single-point failures.

Proof 3: According to Shamir's constraints, it is impossible for t numbers of CC_i to collude simultaneously to recover blind factors and decryption keys. One of these multiple CC is unable to participate in the data aggregation protocol due to a single-point attack or an accident. As long as there are no fewer than t CC_i that can operate normally, work can continue to resume. This meets the contribution requirements for robustness that we previously proposed.

Theorem 2: This scheme has public verifiability to ensure that the control center receives the correct secret shares.

Scenario 4: Attackers' attempts to tamper with secretly shared information are infeasible within a limited time constraint.

Proof 4: During the Shamir sharding process, safety measures were implemented, namely zero-knowledge proofs. GW need to prove that equation $X = g_2^{C(l)}$, $Y = y_c^{C(l)}$ holds true. When verifying, we know $X = \Pi_{l=1}^{t-1} g_2^{d_l} \cdot g_2^C = g_2^{C(l)}$, $w_1 = g_2^r$, $w_2 = y^r$, $u = hash(w_1||X||w_2||Y)$, $v = r + C(l)umodq$, check if GW has encrypted the correct sub shares and verify if the following equation holds: $g_2^v = g_2^{r+C(l)u} = g_2^r \cdot g_2^{C(l)u} = w_1 \cdot X^u$ and $y^v = y^{r+C(l)u} = y^r \cdot y^{C(l)u} = w_2 \cdot Y^u$.

If an adversary attempts to obtain or tamper with a ciphertext fragment, it is necessary to solve the discrete logarithm problem within a transmission duration, which is somewhat impossible. This meets the design goal of verifiability that we proposed earlier.

5.2   Performance Analysis

Experimental setup: The platform for experimenting is a computer with an Intel Core i5 CPU(1.60GHz) running Windows and 8 GB RAM. The OS is Windows 10 Professional Edition. Our programming language is JAVA and IDE is Intellij IDEA 2022.

The operation carried out at the smart grid end is to first receive the blind factor x_i and encryption parameter (N, g) from the trusted center during the initialization phase, and then generate ciphertext c_i and data increment Δm_i through these. Therefore, the calculation amount of each smart meter in one time slot is $2T_e + 2T_d + T_h + T_m = 0.013942$ ms. The communication overhead for transmitting a ciphertext is 2048 bits, and the communication overhead for transmitting additional data increments is 1024 bits. Therefore, the total communication overhead of sending the smart meter to the aggregator is 3072 bits.

The operation required on the aggregator side is to receive ciphertext c_i, Δm_i from n smart meters and aggregate it. Obtain the aggregated value C of

Table 2. Computational costs

Operation	Description	Time consumption
T_e	Exponential operation	0.003600 ms
T_d	Dot multiplication operation	0.000111ms
T_h	Hash operation	0.006400 ms
T_m	Modulo operation	0.000120 ms
T_a	Addition operation	0.000112 ms
T_{share}	Shamir secret sharding operation	0.096400 ms
T_{zero}	Publicly verifiable operations	0.007200 ms

ciphertext and the aggregated value SUM' of data increment, then perform Shamir sharding and perform zero-knowledge proof preprocessing on s shares separately. Therefore, the computation required for each aggregator to perform a protocol is $(n-1)T_d + (n-1)T_a + 2T_{share} + sT_{zero}$. The aggregator transmits $Y = y^{c_i}$ to each control center, resulting in a communication overhead of 524288 bits.

6   Conclusion

This study presents a novel publicly verifiable multi-subset data aggregation scheme designed to bolster the system's robustness and credibility. Leveraging Shamir's secret sharing technology, the scheme encrypts data, divides it into multiple segments, and employs zero-knowledge proof technology to ensure secure transmission to various control centers. Security analysis and experimental results demonstrate the scheme's efficacy in thwarting collusion attacks and mitigating single-point failures, validating its practical application. However, the protocol is not quantum-resistant. Future research will focus on developing quantum-resistant protocols for the Internet of Things environment.

Acknowledgements. This work was supported in part by the National Natural Science Foundation of China (Grant No. 62072133), in part by Wenzhou Science and Technology Plan (Grant No. ZG2023028).

References

1. Bao, H., Lu, R.: A new differentially private data aggregation with fault tolerance for smart grid communications. IEEE Internet Things J. **2**, 248–258 (2015). https://api.semanticscholar.org/CorpusID:18374228
2. Chen, C., Kishore, S., Snyder, L.V.: An innovative RTP-based residential power scheduling scheme for smart grids. In: 2011 IEEE International Conference on Acoustics, Speech and Signal Processing (ICASSP) pp. 5956–5959 (2011). https://api.semanticscholar.org/CorpusID:8285849

3. Chen, D., Zhou, T., Liu, W., Li, R., Wu, L., Yang, X.: MDA-FLH: multidimensional data aggregation scheme with fine-grained linear homomorphism for smart grid. IEEE Internet Things J. **11**(2), 3524–3538 (2024). https://doi.org/10.1109/JIOT.2023.3297734

4. Chen, Y., Yang, S., Martínez-Ortega, J.F., López, L., Yang, Z.: A resilient group-based multisubset data aggregation scheme for smart grid. IEEE Internet Things J. **10**(15), 13649–13661 (2023). https://doi.org/10.1109/JIOT.2023.3262731

5. Fang, X., Misra, S., Xue, G., Yang, D.: Smart grid – the new and improved power grid: a survey. IEEE Commun. Surv. Tutor. **14**, 944–980 (2012). https://api.semanticscholar.org/CorpusID:12103882

6. Kong, W., Dong, Z.Y., Jia, Y., Hill, D.J., Xu, Y., Zhang, Y.: Short-term residential load forecasting based on LSTM recurrent neural network. IEEE Trans. Smart Grid **10**, 841–851 (2019). https://api.semanticscholar.org/CorpusID:57362815

7. Li, S., Xue, K., Yang, Q., Hong, P.: Ppma: Privacy-preserving multisubset data aggregation in smart grid. IEEE Trans. Ind. Inform. **14**, 462–471 (2018). https://api.semanticscholar.org/CorpusID:4651456

8. Liang, Y., Liu, Y., Zhang, X., Liu, G.: Physically secure and privacy-preserving charging authentication framework with data aggregation in vehicle-to-grid networks. IEEE Trans. Intell. Transp. Syst. 1–16 (2024). https://doi.org/10.1109/TITS.2024.3443171

9. Liu, H., Gu, T., Shojafar, M., Alazab, M., Liu, Y.: Opera: optional dimensional privacy-preserving data aggregation for smart healthcare systems. IEEE Trans. Industr. Inf. **19**(1), 857–866 (2023). https://doi.org/10.1109/TII.2022.3192037

10. Lu, R., Alharbi, K., Lin, X., Huang, C.: A novel privacy-preserving set aggregation scheme for smart grid communications. In: 2015 IEEE Global Communications Conference (GLOBECOM), pp. 1–6 (2014). https://api.semanticscholar.org/CorpusID:16397469

11. Ma, W., Liu, X., Yu, J., Yu, K., Wang, X.: A collusion attack resistance data aggregation scheme in internet of things. IEEE Trans. Ind. Inform. 1–11 (2024). https://doi.org/10.1109/TII.2024.3435513

12. Manikandan, J., Srilakshmi, U.: Multi-parameter secure data aggregation in data centre with integrity verification. In: 2023 International Conference on Ambient Intelligence, Knowledge Informatics and Industrial Electronics (AIKIIE), pp. 01–09 (2023). https://doi.org/10.1109/AIKIIE60097.2023.10390507

13. Merad-Boudia, O.R., Senouci, S.M.: An efficient and secure multidimensional data aggregation for fog-computing-based smart grid. IEEE Internet Things J. **8**(8), 6143–6153 (2021). https://doi.org/10.1109/JIOT.2020.3040982

14. Nguyen, N.T., Liu, B.H., Nguyen, N.P., Chou, J.T.: Cyber security of smart grid: attacks and defenses. In: ICC 2020 - 2020 IEEE International Conference on Communications (ICC), pp. 1–6 (2020). https://api.semanticscholar.org/CorpusID:220891699

15. Pang, H., He, K., Fu, Y., Liu, J.N., Liu, X., Tan, W.: Enabling efficient and malicious secure data aggregation in smart grid with false data detection. IEEE Trans. Smart Grid **15**(2), 2203–2213 (2024). https://doi.org/10.1109/TSG.2023.3316730

16. Peng, C., Luo, M., Vijayakumar, P., He, D., Said, O., Tolba, A.: Multifunctional and multidimensional secure data aggregation scheme in WSNs. IEEE Internet Things J. **9**(4), 2657–2668 (2022). https://doi.org/10.1109/JIOT.2021.3077866

17. Peng, C., Luo, M., Wang, H., Khan, M.K., He, D.: An efficient privacy-preserving aggregation scheme for multidimensional data in IotT. IEEE Internet Things J. **9**(1), 589–600 (2022). https://doi.org/10.1109/JIOT.2021.3083136

18. Song, J., Liu, Y., Shao, J., Tang, C.: A dynamic membership data aggregation (DMDA) protocol for smart grid. IEEE Syst. J. **14**, 900–908 (2020). https://api. semanticscholar.org/CorpusID:164582200

19. Wang, C., Shen, J., Vijayakumar, P., Gupta, B.B.: Attribute-based secure data aggregation for isolated IoT-enabled maritime transportation systems. IEEE Trans. Intell. Transp. Syst. **24**(2), 2608–2617 (2023). https://doi.org/10.1109/TITS.2021. 3127436

20. Wang, Y., Chen, Q., Hong, T., Kang, C.: Review of smart meter data analytics: applications, methodologies, and challenges. IEEE Trans. Smart Grid **10**, 3125–3148 (2018). https://api.semanticscholar.org/CorpusID:3620111

21. Yan, X., et al.: Verifiable, reliable, and privacy-preserving data aggregation in fog-assisted mobile crowdsensing. IEEE Internet Things J. **8**(18), 14127–14140 (2021). https://doi.org/10.1109/JIOT.2021.3068490

22. Yang, Q., Zhifan, D.: Privacy-preserving data aggregation for single point of failure in smart grid. In: 2021 24th International Symposium on Wireless Personal Multimedia Communications (WPMC), pp. 1–6 (2021).https://doi.org/10.1109/ WPMC52694.2021.9700450

23. Zeng, Z., Wang, X., Liu, Y.N., Chang, L.: MSDA: multi-subset data aggregation scheme without trusted third party. Front. Comput. Sci. **16** (2021). https://api. semanticscholar.org/CorpusID:238639170

24. Zhang, J., Zhao, Y., Wu, J., Chen, B.: LVPDA: a lightweight and verifiable privacy-preserving data aggregation scheme for edge-enabled IoT. IEEE Internet Things J. **7**(5), 4016–4027 (2020). https://doi.org/10.1109/JIOT.2020.2978286

25. Zhang, J., Wang, Y., Ma, Z., Yang, X., Ying, Z., Ma, J.: A location-aware verifiable outsourcing data aggregation in multiblockchains. IEEE Internet Things J. **10**(6), 4783–4798 (2023). https://doi.org/10.1109/JIOT.2022.3221555

26. Zhao, J., Huang, H., Zhang, X., He, D., Choo, K.K.R., Jiang, Z.L.: VMEMDA: verifiable multidimensional encrypted medical data aggregation scheme for cloud-based wireless body area networks. IEEE Internet Things J. **11**(10), 18647–18662 (2024). https://doi.org/10.1109/JIOT.2024.3365909

Assessing the Effectiveness of LLMs in Android Application Vulnerability Analysis

Vasileios Kouliaridis[1]([envelope]) [iD], Georgios Karopoulos[1] [iD],
and Georgios Kambourakis[2] [iD]

[1] European Commission, Joint Research Centre (JRC), 21027 Ispra, Italy
vasileios.kouliaridis@ec.europa.eu
[2] Department of Information and Communication Systems Engineering,
University of the Aegean, Karlovasi, 83200 Samos, Greece

Abstract. The increasing frequency of attacks on Android applications coupled with the recent popularity of large language models (LLMs) necessitates a comprehensive understanding of the capabilities of the latter in identifying potential vulnerabilities, which is key to mitigate the overall risk. To this end, the work at hand compares the ability of nine state-of-the-art LLMs to detect Android code vulnerabilities listed in the latest Open Worldwide Application Security Project (OWASP) Mobile Top 10. Each LLM was evaluated against an open dataset of over 100 vulnerable code samples, assessing each model's ability to identify key vulnerabilities. Our analysis reveals the strengths and weaknesses of each LLM, identifying important factors that contribute to their performance. Additionally, we offer insights into context augmentation with retrieval-augmented generation (RAG) for detecting Android code vulnerabilities, which in turn may propel secure application development. Finally, while the reported findings regarding code vulnerability analysis show promise, they also reveal significant discrepancies among the different LLMs.

Keywords: Large Language Models · Vulnerability analysis · Code analysis · OWASP · Mobile security · Android · Retrieval-Augmented Generation

1 Introduction

As mobile devices continue to proliferate, the need for secure software development practices remains still of high priority. The predominant Android platform has become a prime target for attackers and malware writers, seeking to exploit vulnerabilities in the vast cosmos of mobile applications [1]. The importance and volume of mobile vulnerabilities has led the Open Web Application Security Project (OWASP) to periodically publish a current, reputable list of the most prevalent vulnerabilities detected in mobile applications, namely OWASP Mobile

W. Meng et al. (Eds.): ADIoT 2024, LNCS 15397, pp. 139–154, 2025.
https://doi.org/10.1007/978-3-031-85593-1_9

Top 10. [2]. This list can serve as a key benchmark in assessing the performance of any tool in finding software vulnerabilities [3].

An emerging approach to detecting Android code vulnerabilities is the use of large language models (LLMs) for code analysis. Actually, the use of LLMs for code analysis is traced back to the early 2010s. That is, in 2013, the introduction of Word2Vec [4], a shallow neural network, marked the beginning of deep learning-based language models. That algorithm was capable of learning word embeddings (an encoding of the meaning of the word) from large datasets. In 2018, Google introduced Word2Vec's successor, a language model known as Bidirectional Encoder Representations from Transformers (BERT) [5]. BERT was designed to be bidirectionally trained, meaning it can learn information from both the left and right sides of a given text during training, therefore obtaining a better understanding of the context.

In the realm of code analysis, LLMs began to gain traction around 2017. One of the early applications of LLMs in code analysis was code completion. Models like GPT-2 [6], fully released in Nov. 2019, were trained on a large corpus of source code data. By understanding the structure and context of the code, these models could predict the most likely code to follow a given input. In 2020, OpenAI [7] introduced GPT-3 [8], a significantly larger model with 175B parameters. This model showed improved capabilities in generating human-like text and was even able to generate code when given a task description. The ability of LLMs to analyze and understand code has also been demonstrated in recent studies [9,10]. Nevertheless, to the best of our knowledge, the literature lacks a comprehensive comparison of the ability of these models to detect Android code vulnerabilities so far.

The present work aims to fill this gap by comparing the ability of nine state-of-the-art LLMs to detect Android code vulnerabilities listed in the OWASP Mobile Top 10. Specifically, each model is evaluated regarding its performance in identifying key vulnerabilities against a dataset comprising snippets of vulnerable Android code. The assessment of each model is done through a combination of manual and automated evaluation methods. We additionally pinpoint the strengths and weaknesses of each LLM and provide insights into the factors that conduce to their performance. Overall, this study provides valuable insights into the use of LLMs for detecting the most common mobile code vulnerabilities, such as those mentioned in OWASP Mobile Top 10. The results of our work can support software engineers in secure mobile app development, integrators in assessing the security of third party software, as well as security evaluators during the review process of mobile app stores. By following our methodology, at least a part of software security evaluations can be automated, decreasing the overhead of manual assessment operations. The contributions of the paper are summarized as follows.

– We present a thorough comparative analysis on the capabilities and performance of nine leading LLMs, i.e., GPT 3.5, GPT 4, GPT 4 Turbo, Llama 2, Zephyr Alpha, Zephyr Beta, Nous Hermes Mixtral, MistralOrca, and Code Llama in identifying vulnerabilities residing in Android applications. The

experiments conducted provide concrete evidence of the LLMs' capabilities
for such tasks, also identifying the limitations per LLM. These insights are
critical for anyone interested in understanding the trade-offs associated with
each LLM.

– We provide a comparison between the code analysis results as given by the
  nine LLMs against two well-known, publicly available static application security testing (SAST) tools, namely, Bearer [11] and MobSFscan [12].
– We examine the impact of context augmentation on LLMs and contribute
  a set of guidelines regarding the selection and fine-tuning of LLMs towards
  enhancing the security posture of Android code.
– We offer an open dataset to the community for driving research in this field
  forward.

The remainder of this paper is structured as follows. The next section presents
previous work on the use of LLM for code vulnerability analysis. Section 3 details
our methodology, while the results per LLM are given in Sect. 4. The last section
concludes and proposes some lines for future research.

2   Previous Work

In recent years, LLMs have gained significant attention in the field of cybersecurity for their potential to provide assistance in various domains, including
vulnerability detection, penetration testing, and security analysis. State-of-the-
art surveys such as [13–15], as well as a more recent but not yet peer-reviewed
study [16], provide comprehensive overviews of the current state and potential
future applications of LLMs in cybersecurity. These works analyze the challenges,
practical implications, and future research directions to exploit the full potential
of these models in ensuring cyber resilience. The rest of this section will focus on
literature dealing with software vulnerability analysis using LLMs. This includes
works that have already been peer-reviewed, as well as more recent research that
has been self-archived for the sake of completeness.

In [17], transformer-based LLMs are evaluated in the task of code vulnerability detection. The authors evaluate such LLMs, including BERT, DistilBERT,
CodeBERT, GPT-2 and Megatron, against C/C++ source code snippets from
two publicly available datasets. The results showed that LLMs perform well in
software vulnerability tasks; indicatively, the best scoring model, GPT-2, had
an F1-score above 95% in all tests. In the context of software engineering, [18]
investigates the use of in-context learning to improve the ability of LLMs to
detect software vulnerabilities, showcasing the adaptability of LLMs to learn
from context-specific examples. The authors use code retrieval to search for
code snippets that are similar to the examined code and feed them to the LLM
together with the examined code and its analysis. Their experimental results
show that this approach has better performance than the original GPT model.

Another set of works, adds verification in the vulnerability detection process. An empirical study of using LLMs for vulnerability assessment in software

was conducted in [19]. The authors used four well-known pre-trained LLMs to identify vulnerabilities in two labeled datasets, namely code gadgets and CVE-fixes, and static analysis as a reference point. The used LLMs include GPT-3.5, Davinci and CodeGen, and the analysis was limited to two kinds of vulnerabilities: SQL injections and buffer overflows. The study concluded that LLMs do not perform well at detecting vulnerabilities, presenting high false-positive rates, but could complement and improve the traditional static analysis process. Concerns about the safe use of code assistants are addressed in [20]. In this case, LLMs are used to produce code which is then assessed manually and using static analysis. This study provides empirical insights into how developers interact with LLMs, underscoring the importance of user awareness to mitigate security risks associated with assisted code generation.

Moving to non peer-reviewed works, the work of [21] delves into the application of LLMs in static binary taint analysis, demonstrating how these models can assist in vulnerability inspection of binaries. A binary is first disassembled and decompiled, and an LLM is used to identify security sensitive functions that may contain vulnerabilities, as well as candidate dangerous flows. In the last phase, the LLM combines the previous results to produce a vulnerability report for the examined binary. The authors of [22] propose DefectHunter, a vulnerability detection mechanism that combines various technologies, including LLMs. Its architecture has three main building blocks: a tool for extracting structural information from code snippets, a pre-trained LLM for generating semantic information, and a Conformer mechanism to identify vulnerabilities from the previously extracted structural and semantic data.

The authors of [23] evaluated ChatGPT and GPT-3 in detection of Common Weakness Enumeration (CWE) vulnerabilities contained in code. Using a custom real-world dataset with Java files from open GitHub repositories, they concluded that the detection capabilities of the aforementioned models are limited. In [24], an empirical study of the potential of LLMs for detecting software vulnerabilities is presented. The authors tested 129 code samples from various GitHub repositories, written in eight different languages, and their results showed that GPT-4 identified around four times more vulnerabilities than traditional, rule-based, static code analysis tools. In addition, the LLMs were asked to provide fixes for the identified vulnerabilities. The models used include GPT-3 and GPT-4.

Apart from generic code, LLMs have been used for detecting vulnerabilities in smart contracts. LLM4Vuln [25] is an evaluation framework for vulnerability detection systems based on LLMs, focusing on smart contract vulnerabilities. The difference from other similar works is that, instead of benchmarking the performance of LLMs in vulnerability detection, the authors evaluate the vulnerability reasoning capabilities of each model. Similarly, the authors of [26] proposed GPTLens, a framework for detecting vulnerabilities in smart contracts using LLMs. GPTLens takes a different approach from the traditional one-stage detection in order to decrease false positives. The detection process is broken down in two steps, where the LLM takes two different roles: auditor and critic. As an auditor, the LLM provides a large range of vulnerabilities for the examined contract, whereas as a critic it verifies the claims produced in the first step. The performed experiments show that GPTLens presents improved results over the single-stage vulnerability detection.

3 Methodology

This section details our methodology, including the creation of the benchmark dataset, the selection of LLMs, and the evaluation process.

3.1 Dataset

Also with reference to Sect. 2, to our knowledge, there is no publicly available dataset containing vulnerable Android code covering each one of the OWASP Mobile Top 10 vulnerabilities. The most relevant dataset to our study is LVDAndro [27], which however is labelled based on CWE. Additionally, since LVDAndro was created using actual Android applications, it contains a significant proportion of non-vulnerable code. In view of this shortage, for the needs of our experiments, we created a new dataset coined *Vulcorpus* [28] containing 100 pieces of vulnerable code. It is important to note that the term "piece of code", hereafter called *sample*, refers to a part of an application, not its full codebase. All the samples were written in Java by exploiting common insecure coding practices, e.g., using weak authentication mechanisms, not filtering input/objects, etc., and target the Android OS. However, obviously, the same vulnerabilities apply to other mobile platforms, say, iOS.

More specifically, Vulcorpus contains 10 samples for each of the OWASP Mobile Top-10 vulnerabilities of 2024, which are briefly explained in Subsect. 3.2. Every sample exhibits maximum two interrelated vulnerabilities, while one or two of these samples per vulnerability category are obfuscated manually using the well-known rename technique. Moreover, to assess each LLM in detecting privacy-invasive code, we created three more samples which perform risky actions without asking the user for confirmation. These actions are:

– Get the precise location of the device through the "android.permission.ACCESS_FINE_LOCATION" permission, and directly share the latitude and longitude over the Internet via API. According to the Android API [29], this permission has a "dangerous" protection level, namely it may give the requesting application access to user's private data, among others.
– Capture an image via the "ACTION_IMAGE_CAPTURE" intent [30] , and subsequently attempt to share the captured image file via API.
– Open local documents through the "ACTION_OPEN_DOCUMENT" intent [31], and attempt to send them to a remote host via API.

The latter three samples are also available at [28] along with Vulcorpus.

3.2 List of Vulnerabilities

This subsection briefly delineates each vulnerability contained in the current OWASP Mobile Top 10 list. For more details regarding each vulnerability, the reader is referred to [2]. It is important to note that the list differs from its 2016 version, given that four vulnerabilities contained in the 2016 list have been replaced with new ones in the current list. The reader should also keep in mind that while some categories of vulnerabilities, say, M5 are straightforward, others might be more complicated for LLMs to understand, such as the M7.

Improper Credential Usage (M1): Poor credential management can lead to severe security issues, namely, unauthorized users may be able to gain access to sensitive information or administrative functionalities within the mobile app or its backend systems. This in turn leads to data breaches and fraudulent activities.

Inadequate Supply Chain Security (M2): By exploiting vulnerabilities in the mobile supply chain, attackers may be able to manipulate application functionality. For example, they can insert malicious code into the mobile application's codebase or libraries [32], as well as modify the code during the application's build process to introduce backdoors, spyware, or other type of malware. The attacker can also exploit vulnerabilities in third-party software libraries, software development kits (SDKs), or hard-coded credentials to gain access to the mobile app or the backend servers. Overall, this type of vulnerabilities can lead to unauthorized data access or manipulation, denial of service, or complete takeover of the mobile application or device.

Insecure Authentication/Authorization (M3): Poor authorization could lead to the destruction of systems or unauthorized access to sensitive information, while poor authentication results in the inability to identify the user making an action request, leading to the inability to log or audit user activity. This situation makes it difficult to detect the source of an attack, understand any underlying exploits, or develop strategies to prevent future attacks. Obviously, authentication failures are tightly coupled to authorization failures; when authentication controls fail, authorization cannot be performed. That is, if an attacker can anonymously execute sensitive functionality, it indicates that the underlying code is not verifying the user's permissions, highlighting failures in both authentication and authorization controls.

Insufficient Input/Output Validation (M4): A mobile application that does not adequately validate and sanitize data from external sources, like user inputs or network data, is susceptible to a range of attacks, including SQL injection, command injection, and cross-site scripting. Insufficient output validation can also lead to data corruption or presentation vulnerabilities, possibly allowing the malicious actor to inject harmful code or manipulate sensitive information shown to the users.

Insecure Communication (M5): Modern mobile applications typically communicate with one or more remote servers. This renders user data susceptible to interception and modification, if they are transmitted in plaintext or using an outdated encryption protocol.

Inadequate Privacy Controls (M6): Privacy controls aim to safeguard Personally Identifiable Information (PII), including names and addresses, credit card details, emails, and information related to health, religion, sexuality, and political opinions. This sensitive information can be used to impersonate the victim for fraudulent activities, misuse their payment data, blackmail them with sensitive information, or harm them by destroying or manipulating sensitive data.

Insufficient Binary Protections (M7): The application's binary may hold valuable information, such as commercial API keys or hard-coded cryptographic secrets. Furthermore, the code within the binary itself could be valuable, for instance, containing critical business logic or pre-trained AI models. In addition to gathering information, attackers may also manipulate app binaries to gain access to paid features for free or to bypass other security controls. In the worst-case scenario, popular apps could be altered to include malicious code and then distributed through third-party app stores or under a new name to deceive unsuspecting users.

Security Misconfiguration (M8): These occur when security settings, permissions, or controls are improperly configured, leading to vulnerabilities and unauthorized access.

Insecure Data Storage (M9): Such vulnerabilities may stem from weak encryption, insufficient data protection, insecure data storage mechanisms, and improper handling of user credentials.

Insufficient Cryptography (M10): The use of obsolete cryptographic suites, primitives, or cryptographic practices may lead to loss of data confidentiality, integrity, and inability to impose source authentication among others. Typical repercussions include data decryption, manipulation of cryptographic processes, leak of encryption keys, etc.

3.3   Selection of LLM

For the purposes of our experiments, nine contemporary, well-known LLMs were chosen: three commercial models, i.e., GPT-3.5, GPT-4, and GPT-4 Turbo, and six open source models, i.e., Llama 2, Zephyr Alpha, Zephyr Beta, Nous Hermes Mixtral, MistralOrca, and Code Llama. According to their documentation, these models have been pre-trained on large amounts of text data, including code, having demonstrated performance in various software engineering tasks, including code analysis. That is, their ability to understand code syntax and semantics makes them well-suited for identifying vulnerabilities residing in code. Additionally, their large size and diverse training data make them less likely to overfit to a specific codebase. A succinct description of each LLM is given below. It is important to note that we used the default settings of each model.

– GPT 3.5 (version 1106) [8]: It is a powerful language model that has been pre-trained on a large corpus of text data, including code. It has demonstrated performance in various natural language processing (NLP) tasks and has been used for code analysis tasks such as code completion, code search, and code summarization.

- GPT 4 [33,34] (version gpt-4-32k): It is the newest version of GPT, pre-trained on an even larger corpus of text data, including code. It has demonstrated improved performance over GPT 3.5 in various NLP tasks and has been used for code analysis, including code review and repair.
- GPT 4 Turbo (version turbo-2024-04-09): It is a variant of GPT 4, specifically designed for tasks that require faster inference times, such as code analysis. It has been pre-trained on the same large corpus of text data as GPT 4, optimized for faster performance.
- Llama 2 (version Llama-2-70b-chat) [35]: This LLM has been pre-trained on a diverse set of text data, including code. It has demonstrated performance in various NLP tasks, also been exploited for code analysis, including code summarization and code search.
- Zephyr Alpha (version zephyr-7b-alpha) [36]: It is pre-trained on a huge corpus of text data from diverse sources, including books, articles, and websites. This model has been fine-tuned with a mix of publicly available and synthetic datasets on top of Mistral LLM. Despite its small size (7B parameters), it potentially shows a performance comparable to several models with a number of parameters in the range of 20-30B.
- Zephyr Beta (version zephyr-7b-beta) [36]: This model has been fine-tuned with a mix of publicly available and synthetic datasets on top of Mistral LLM. It is the successor of Zephyr Alpha, therefore considered significantly more powerful than its predecessor. Based on its documentation, it is fast and competent, showing a performance comparable to the best open-source models, having around 70B parameters.
- Nous Hermes Mixtral (version nous-hermes-2-mixtral-8x7b-dpo) [37]: It is one of the most powerful open-source models available, comprising a fine-tuned version of Mixtral base model.
- MistralOrca (version mistral-7b-openorca [38–40]: It has been fine-tuned with Open-Orca datasets on top of Mistral LLM. Despite its small size, it outperforms Llama 2 13B, showing a performance comparable to several models with a number of parameters in the range of 20-30B.
- Code Llama [41] (version 7b): It is a special version of Llama 2, tailored specifically for coding applications. This specialized version has been refined through extensive additional training on code-focused data, with prolonged exposure to relevant datasets. The result is a tool with alleged superior coding proficiency that builds upon the foundation of Llama 2. More specifically, Code Llama can generate code and create explanations about code in response to prompts in both programming and natural language. Its capabilities extend to assisting with code completion and troubleshooting code errors. Furthermore, Code Llama is versatile, supporting a broad array of widely-used programming languages, including Python, C++, Java, PHP, JavaScript, C Sharp, and Bash. In this work, we examine the smallest pre-trained model, namely, the 7B version. In addition, for this LLM, in a separate run, we employed LlamaIndex [42] to improve the detection capabilities of Code Llama. LlamaIndex is a data framework for LLM-based applications, enhancing them with additional contextual data. This context augmenta-

tion technique is called Retrieval-Augmented Generation (RAG) and can be used to address the restrictions of LLMs by giving them access to contextual, current data. For the RAG process, we selected the bge-small-en-v1.5 [43] embedding model, developed by the Beijing Academy of Artificial Intelligence. Additionally, we used the 50% of Vulcorpus, i.e., only the samples that contain code comments regarding the specific vulnerability. Android's application quality and security guidelines and code examples [44] were also added as input to the RAG, along with information on each vulnerability from the OWASP website [2].

3.4   Evaluation Process

All nine pre-trained LLMs listed in Subsect. 3.3, except Code Llama, run on the *GPT@JRC* platform, a system developed by the European Commission's Joint Research Centre (JRC). Code Llama was run on a local computer with an M2 processor and 16 GB unified memory. Each LLM was fed with Vulcorpus for comparing its performance on identifying potential vulnerabilities and proposing code improvements. To this end, as detailed in Sect. 4, we use a simple scoring system to present (a) the number of vulnerabilities each LLM was able to detect, and (b) if the LLM proposed valid suggestions for possibly fixing the vulnerability. Both these partial scores have a maximum value of 10/10 per vulnerability category, i.e., one point for each piece of vulnerable code the LLM was able to detect and annotate. It is important to note that the input or question given to each LLM has a major effect on its output. For our study, each LLM was queried as follows: "Check if there are any security issues in the following code; if there are, explain the issue".

As previously mentioned, the LLMs used in this work are pre-trained. This means that the associated libraries, possibly needed by each code sample but not included in the input, cannot be analyzed. This mostly affects the analysis regarding the M2 vulnerability. Therefore, to evaluate LLMs against M2, instead of Java code, we used 10 libraries with known vulnerabilities as input. These libraries, also included in Vulcorpus for reasons of reproducibility, were published before the training date of each LLM.

At a final stage, as detailed in Sect. 4, the results of each LLM were compared and crosschecked against those produced by two well-known SAST tools, namely Bearer [11] and MobSFscan [12]. Bearer is a static application security testing tool, which uses built-in rules covering the OWASP Top 10 and CWE Top 25. MobSFscan is a static analysis tool that uses MobSF's [45] security rules and can find insecure code patterns in Android or iOS source code. Finally, we also assessed the performance of each LLM in detecting privacy-invasive behaviors, using the three samples detailed in Subsect. 3.1. The output was rated using three categories: (a) not privacy-invasive, (b) potentially privacy-invasive, and (c) privacy-invasive.

4   Results

Tables 1 and 2 recapitulate the results for each LLM. Particularly, each line of Table 1 indicates if the specific model *Detected* the vulnerability (denoted with the letter "D"), and if it explained the situation and provided a valid solution for *Improving* the code (denoted with the letter "I"). Actually, the "I" aspect is a key factor in evaluating each LLM (also against each other), as this is the sole indicator of whether the LLM actually "perceives" the security issue.

Table 1. Vulnerability analysis results. The letters "D" and "I" stand for the number of vulnerable samples detected and the number of vulnerable samples for which the LLM suggested improvements, respectively. Top scores per vulnerability are in boldface. The asterisk exhibitor stands for Code Llama without RAG.

LLM	M1		M2		M3		M4		M5		M6		M7		M8		M9		M10		Mean	
	D	I	D	I	D	I	D	I	D	I	D	I	D	I	D	I	D	I	D	I	D	I
GPT-3.5	3	3	**7**	N/A	2	3	2	3	8	6	3	5	5	4	3	5	4	6	5	2	4.2	4.1
GPT-4	**10**	**10**	0	N/A	6	7	6	8	5	**10**	**10**	**10**	6	9	7	**10**	8	**10**	9	9	6.7	**9.2**
GPT-4 TURBO	4	5	0	N/A	3	5	5	8	8	9	6	8	4	4	7	**10**	7	9	6	8	5	7.3
Nous Hermes Mixtral	1	3	6	N/A	1	3	**9**	5	6	8	8	8	7	3	9	9	8	**10**	7	7	6.2	6.2
Mistral Orca	9	9	0	N/A	2	2	4	4	5	5	0	0	3	4	0	1	**10**	**10**	4	3	3.7	4.2
Zephyr Alpha	0	0	3	N/A	6	6	5	5	**10**	**10**	7	8	2	2	7	7	3	**10**	**10**	**10**	5.3	6.4
Zephyr Beta	0	8	0	N/A	9	9	8	8	**10**	**10**	0	8	3	0	5	4	**10**	0	9	9	5.4	6.2
Llama 2	0	0	0	N/A	0	**10**	4	5	**10**	0	6	6	0	0	0	0	4	4	6	6	3	3.4
Code Llama*	9	5	3	N/A	9	4	8	4	**10**	5	8	5	9	4	9	4	7	7	9	6	**8.1**	4.9

Overall, with reference to Table 1, the best performers in terms of total vulnerabilities detected, are Code Llama (81/100), GPT 4 (67/100), Nous Hermes Mixtral (62/100), Zephyr Beta (54/100), and Zephyr Alpha (53/100), followed by GPT 4 TURBO (50/100), GPT 3.5 (42/100), MistralOrca (37/100), and Llama 2 (30/100). On the other hand, the best performers, in terms of total code improvement suggestions, are GPT 4 (83/90), GPT 4 Turbo (66/90), Zephyr Alpha (58/90), Zephyr Beta (56/90), and Nous Hermes Mixtral (56/90), followed by Code Llama (44/90), MistralOrca (38/90), GPT 3.5 (37/90), and Llama 2 (31/90). Overall, GPT 4 poses as the top performer, considering a composite score of high "D" and high "I". On the other hand, LLMs like Code Llama, which do identify the correct vulnerability, but fail to provide corrections or suggestions regarding the problematic lines of code may indicate an insufficiently trained model for this type of analysis.

When looking at each vulnerability individually, GPT 4 achieved a perfect score for M1 and M6, MistralOrca for M9, Zephyr alpha for M5 and M10, Zephyr beta for M5 and M9, and Llama 2 and Code Llama for M5. Regarding the rest of the vulnerabilities, namely, M2, M3, M4, M7, and M8, the best performers were

Table 2. Results per LLM regarding privacy-invasive actions. N: not privacy-invasive, P: potentially privacy-invasive, Y: privacy-invasive.

LLM	Location	Camera	Files
GPT 3.5	Y	P	P
GPT 4	N	P	N
GPT 4 Turbo	P	Y	P
Nous Hermes Mixtral	Y	P	P
MistralOrca	N	N	N
Zephyr Alpha	Y	P	Y
Zephyr Beta	Y	P	N
Llama 2	P	Y	N
Code Llama	N	N	Y
Code Llama + RAG	N	Y	P

GPT 3.5 (7/10), Zephyr Beta and Code Llama (9/10), Nous Hermes Mixtral (9/10), Code Llama (9/10), and Nous Hermes Mixtral and Code Llama (9/10), respectively. Concerning M2, recall from Subsect. 3.4 that it was tested using 10 vulnerable libraries published before the training date of each LLM. Even so, the M2 low detection performance in Table 1 for all the LLMs but GPT-3.5 may designate that these libraries were not considered during LLM training, so the respective scores can be regarded only as indicative. The same applies to the "I" score for M2, which it is marked as *N/A*. As discussed in Subsect. 3.3, to address these limitations, LLMs used for vulnerability detection can capitalize on context augmentation; this way the LLM is provided with access to contextual, up-to-date data.

After averaging the "D" score for all the nine LLMs, we sort in ascending order the OWASP Top 10 vulnerabilities in Table 3. This mean score provides an estimation of the detection difficulty per vulnerability as experienced by the different LLMs. The same table also includes the best performer(s) along with its score in parentheses. As observed from the table, from an LLM viewpoint, M2 is the toughest vulnerability with an average score of 2.11. As explained in Subsect. 3.4, this poor outcome is conceivably due to lack of sufficient, up-to-date information at the LLMs' side. Generally, this low score is somewhat expected, as for this vulnerability the LLMs are checking for known security issues in a list of libraries instead of analyzing the application's code. On the other hand, the highest average detection score was observed in M5, where four LLMs achieved a perfect score.

No less important, with reference to the last stage of the experiments as given in Subsect. 3.4, regarding the detection of privacy-invasive actions, six, eight, and six of the LLMs correctly perceived potential privacy-invasive actions for location, camera, and local file sharing, respectively. The best performer was Zephyr Alpha, which clearly marked two out of three codes as privacy-invasive

Table 3. Best performing LLM model and average detection score per vulnerability. The asterisk exhibitor stands for Code Llama without RAG.

Vulnerability	Best performer	Mean score
M2	GPT 3.5 (7)	2.1
M1	GPT 4 (10)	4
M7	Code Llama* (9)	4.33
M3	Zephyr Beta (9), Code Llama* (9)	4.33
M8	Nous Hermes Mixtral (9), Code Llama* (9)	5.22
M6	GPT 4 (10)	5.33
M4	Nous Hermes Mixtral (9)	5.66
M9	MistralOrca (10), Zephyr Beta (10)	6.77
M10	Zephyr Alpha (10)	7.22
M5	Zephyr Alpha (10), Zephyr Beta (10), Llama 2 (10), Code Llama* (10)	8

and the other as potentially privacy-invasive. The worst performer in this type of experiments was MistralOrca, which was unable to detect any possible privacy-invasive actions.

Additionally, Table 4 presents the results regarding the use of RAG on Code Llama. As explained in Subsect. 3.3, in this experiment, only half of the samples per vulnerability were indexed for RAG, along with text and code examples from Android's app security guidelines [44] and all the CVEs related to the vulnerable libraries used for M2. The samples indexed for RAG were annotated with comments explaining the vulnerable code. After that, we analyzed the other half of the samples, i.e., the non-annotated ones with comments on the particular vulnerability. As observed from Table 4, the results show improvements in both the detection performance and the generation of code suggestions vis-à-vis the base model. Precisely, by feeding a large list of vulnerable libraries, the optimized Code Llama model achieved a perfect score for M2, an improvement of approximately 233% compared to that in Table 1. Nevertheless, for reaching this performance in real-world scenarios, the RAG process should involve an up-to-date dataset comprising known vulnerable library versions. Interestingly, except M2, the optimized Code Llama model detected the vulnerabilities and suggested improvements for all the M1, M3, M4, M5, M6, and M7 samples. As seen in the three bottom lines of Table 4, a nearly perfect performance (4/5) was also observed for all the M8, M9, and M10 samples.

As previously mentioned, the performance of the LLMs was also compared against two reputable SASTs, namely Bearer and MobSFscan. Precisely, as shown in Table 5, across the 100 samples of Vulcorpus, Bearer found 29 security issues, while MobSFscan detected 12 issues. Excluding M2, this result suggests that, for several vulnerability types, the performance of at least some of the LLMs may significantly or even by far surpass that of well-known SASTs. For instance, comparing the numbers of Table 5 with the average scores of Table 3 it can be argued that the former observation applies especially to M3, M4, and M9, and in a smaller extent to M1, M6, and M7. Moreover, a side conclusion is

Table 4. Evaluation results for Code Llama with RAG. The letters "D" and "I" stand for the number of vulnerable samples detected and the number of vulnerable samples for which the LLM suggested improvements, respectively.

Vulnerability	D	I
M1	5/5	5/5
M2	10/10	N/A
M3	5/5	5/5
M4	5/5	5/5
M5	5/5	5/5
M6	5/5	5/5
M7	5/5	5/5
M8	4/5	4/5
M9	4/5	5/5
M10	4/5	5/5

that both the LLMs and SASTs score well in certain vulnerabilities, i.e., M10, and to a lesser extent M5; nevertheless, this is somewhat expected given that vulnerabilities of these two types are generally considered easier to detect.

Table 5. Results of prominent SASTs

SAST	M1	M2	M3	M4	M5	M6	M7	M8	M9	M10	Total
Bearer	2	N/A	0	1	6	3	3	4	3	7	**29**
MobSFscan	2	N/A	0	0	1	1	1	3	0	4	**12**

5 Conclusions

Our study provides empirical evidence regarding the effectiveness of using LLMs for Android code vulnerability analysis. GPT-4 and Code Llama emerged as the top performers among the nine LLMs tested, the latter excelling in detection, but failing to provide sufficient code improvements, and the former showing promising results both in detection and code improvement. Notably, the study highlights the superior performance of specific LLMs for particular types of vulnerabilities. For instance, MistralOrca and Zephyr Beta performed exceptionally well for M9, while Zephyr Alpha excelled in M10. These findings suggest that while some LLMs have a general proficiency in vulnerability detection, others may be more specialized, indicating the potential for strategic selection of LLMs based on the targeted vulnerability type. When comparing open LLM models with commercial ones, we can see that the open models were the best performers in seven out of ten categories of vulnerabilities, i.e., M3, M4, M5, M7, M8, M9, M10. On the other hand, considering mean detection and improvements scores, as presented in Table 1, the situation is mixed.

Our findings also reveal that while some LLMs are capable of detecting Android code vulnerabilities, their overall performance is still in an early stage. For example, several LLMs struggled with M7, while others were unable to identify M2, reflecting the inherent complexity and subtlety of such vulnerabilities. This outcome points to a need for further research towards enhancing LLMs' capabilities in more nuanced areas of Android security. As an additional step, we evaluated the use of RAG in fine-tuning LLMs for vulnerability analysis, with our results demonstrating that RAG can significantly reinforce detection performance. Regarding the detection of privacy-invasive actions, the obtained results indicate a mixed level of sensitivity among the LLMs, with Zephyr Alpha being the top performer. However, MistralOrca's inability to identify any potential privacy-invasive actions underscores the variability in performance and the need for increased model robustness in privacy analysis concerning mobile platforms. No less important, after comparing the performance of LLMs with that of well-respected SASTs on the same set of vulnerable samples, it can be said that the former seem more adept at identifying code vulnerabilities.

Altogether, the results of the present study provide valuable insights into the current state of LLMs in Android vulnerability detection. While certain models show high efficacy, there is ample room for improvement and targeted optimizations, particularly in addressing complex and subtle vulnerabilities. Nevertheless, for obtaining a more complete view, more experiments with larger datasets are needed.

References

1. Lookout: Mobile Threat Landscape Report: 2023 in Review (2023). https://www.lookout.com/threat-intelligence/report/mobile-landscape-threat-report. Accessed 20 Aug 2024
2. OWASP: Mobile Top 10 2024: Final Release Updates (2024). https://owasp.org/www-project-mobile-top-10/. Accessed 20 Aug 2024
3. Kouliaridis, V., Karopoulos, G., Kambourakis, G.: Assessing the security and privacy of android official id wallet apps. Information **14**(8), 457 (2023)
4. Kenneth Ward Church: Word2vec. Nat. Lang. Eng. **23**(1), 155–162 (2017)
5. Devlin, J., Chang, M.-W., Lee, K., Toutanova, K.: BERT: pre-training of deep bidirectional transformers for language understanding (2019)
6. Solaiman, I., et al.: Release strategies and the social impacts of language models (2019)
7. OpenAI: OpenAI (2024). https://openai.com/. Accessed 20 Aug 2024
8. Brown, T.B., et al.: Language models are few-shot learners (2020)
9. Wan, Y., Zhao, W., Zhang, H., Sui, Y., Xu, G., Jin, H.: What do they capture? A structural analysis of pre-trained language models for source code. In: Proceedings of the 44th International Conference on Software Engineering, ICSE 2022, pp. 2377–2388. Association for Computing Machinery, New York (2022)

10. Liu, J., Xia, C.S., Wang, Y., Zhang, L.: Is your code generated by chatGPT really correct? Rigorous evaluation of large language models for code generation (2023)
11. Bearer. https://github.com/Bearer/bearer. Accessed 20 Aug 2024
12. MobSFscan. https://github.com/MobSF/mobsfscan. Accessed 20 Aug 2024
13. Al-Hawawreh, M., Aljuhani, A., Jararweh, Y.: ChatGPT for cybersecurity: practical applications, challenges, and future directions. Clust. Comput. **26**, 3421–3436 (2023)
14. Yao, Y., Duan, J., Kaidi, X., Cai, Y., Sun, Z., Zhang, Y.: A survey on large language model (LLM) security and privacy: the good, the bad, and the ugly. High-Confid. Comput. **4**(2), 100211 (2024)
15. Gupta, M., Akiri, C., Aryal, K., Parker, E., Praharaj, L.: From chatGPT to threatGPT: impact of generative AI in cybersecurity and privacy. IEEE Access **11**, 80218–80245 (2023)
16. Motlagh, F.N., Hajizadeh, M., Majd, M., Najafi, P., Cheng, F., Meinel, C.: Large language models in cybersecurity: State-of-the-art (2024)
17. Thapa, C., Jang, S.I., Ahmed, M.E., Camtepe, S., Pieprzyk, J., Nepal, S.: Transformer-based language models for software vulnerability detection. In: Proceedings of the 38th Annual Computer Security Applications Conference, ACSAC 2022, pp. 481–496. Association for Computing Machinery, New York (2022)
18. Liu, Z., Liao, Q., Gu, W., Gao, C.: Software vulnerability detection with GPT and in-context learning. In: 2023 8th International Conference on Data Science in Cyberspace (DSC), pp. 229–236 (2023)
19. Purba, M.D., Ghosh, A., Radford, B.J., Chu, B.: Software vulnerability detection using large language models. In: 2023 IEEE 34th International Symposium on Software Reliability Engineering Workshops (ISSREW), pp. 112–119 (2023)
20. Sandoval, G., Pearce, H., Nys, T., Karri, R., Garg, S., Dolan-Gavitt, B.: Lost at C: a user study on the security implications of large language model code assistants. In: 32nd USENIX Security Symposium (USENIX Security 2023), Anaheim, CA, August 2023, pp. 2205–2222. USENIX Association (2023)
21. Liu, P., et al.: Harnessing the power of LLM to support binary taint analysis (2023)
22. Wang, J., Huang, Z., Liu, H., Yang, N., Xiao, Y.: DefectHunter: a novel LLM-driven boosted-conformer-based code vulnerability detection mechanism (2023)
23. Cheshkov, A., Zadorozhny, P., Levichev, R.: Evaluation of chatGPT model for vulnerability detection (2023)
24. Noever, D.: Can large language models find and fix vulnerable software? (2023)
25. Sun, Y., et al.: LLM4Vuln: a unified evaluation framework for decoupling and enhancing LLMs' vulnerability reasoning (2024)
26. Hu, S., Huang, T., Ilhan, F., Tekin, S., Liu, L.: Large language model-powered smart contract vulnerability detection: new perspectives. In: 2023 5th IEEE International Conference on Trust, Privacy and Security in Intelligent Systems and Applications (TPS-ISA), Los Alamitos, CA, USA, November 2023, pp. 297–306. IEEE Computer Society (2023)
27. Senanayake, J., Kalutarage, H., Al-Kadri, M.O., Piras, L., Petrovski, A.: Labelled vulnerability dataset on android source code (LVDAndro) to develop AI-based code vulnerability detection models. In: Proceedings of the 20th International Conference on Security and Cryptography - SECRYPT, pp. 659–666. INSTICC. SciTePress (2023)
28. Vulcorpus-2024. https://github.com/billkoul/vulcorpus-2024. Accessed 20 Aug 2024

29. Android developers - manifest permissions. https://developer.android.com/reference/android/Manifest.permission#ACCESS_FINE_LOCATION. Accessed 20 Aug 2024

30. Android developers - mediastore. https://developer.android.com/reference/android/provider/MediaStore#ACTION_IMAGE_CAPTURE. Accessed 20 Aug 2024

31. Android developers - intent. https://developer.android.com/reference/android/content/Intent#ACTION_OPEN_DOCUMENT. Accessed 20 Aug 2024

32. What we know about the XZ utils backdoor that almost infected the world. https://arstechnica.com/security/2024/04/what-we-know-about-the-xz-utils-backdoor-that-almost-infected-the-world/. Accessed 17 Apr 2024

33. OpenAI: GPT-4 is OpenAI's most advanced system, producing safer and more useful responses (2024). Accessed 20 Aug 2024

34. OpenAI: GPT-4 technical report (2024)

35. Touvron, H., et al.: Llama 2: open foundation and fine-tuned chat models (2023)

36. Tunstall, L., et al.: Zephyr: direct distillation of LM alignment (2023)

37. Nous Hermes 2 mixtral 8X7B DPO. https://huggingface.co/NousResearch/Nous-Hermes-2-Mixtral-8x7B-DPO. Accessed 20 Aug 2024

38. Lian, W., et al.: MistralOrca: Mistral-7B model instruct-tuned on filtered openOrcaV1 GPT-4 dataset (2023). Accessed 20 Aug 2024

39. Mukherjee, S., Mitra, A., Jawahar, G., Agarwal, S., Palangi, H., Awadallah, A.: Orca: progressive learning from complex explanation traces of GPT-4 (2023)

40. Longpre, S., et al.: The flan collection: designing data and methods for effective instruction tuning (2023)

41. Code llama. https://ai.meta.com/blog/code-llama-large-language-model-coding/. Accessed 20 Aug 2024

42. Llamaindex. https://docs.llamaindex.ai/en/stable/. Accessed 20 Aug 2024

43. BAAI/bge-small-en-v1.5. https://huggingface.co/BAAI/bge-small-en-v1.5. Accessed 20 Aug 2024

44. Security guidelines. https://developer.android.com/privacy-and-security/security-tips. Accessed 20 Aug 2024

45. MobSF. https://github.com/MobSF/Mobile-Security-Framework-MobSF. Accessed 20 Aug 2024

Singularization: A New Approach to Designing Block Ciphers for Resource-Constrained Devices

Gilles Macario-Rat[1] and Mihail-Iulian Plesa[2(✉)]

[1] Orange, Chatillon, France
gilles.macariorat@orange.com
[2] Orange Services, Bucharest, Romania
mihail.plesa@orange.com

Abstract. Running traditional symmetric encryption algorithms, such as AES, on resource-constrained devices presents significant challenges due to the limited computational resources available. A common bottleneck in these algorithms is the number of rounds, which is typically determined through cryptanalysis efforts. In this paper, we introduce a novel framework for designing block ciphers, termed Singularization. This framework is based on a generic Feistel network with dynamically generated pseudorandom functions (PRFs). We demonstrate that Singularization may enable the design of symmetric ciphers with fewer rounds without compromising security. This is evidenced by a case study of a 6-round DES, which is vulnerable to differential cryptanalysis attacks. By redesigning DES using our framework, we mitigate this vulnerability, suggesting that it is possible to achieve almost the same level of security as a full-round DES with a reduced number of rounds.

Keywords: Cryptography · Moving Target Defense · Block Cipher

1 Introduction

Conventional algorithms, such as AES, introduce significant overhead when executed on resource-constrained devices [2]. To address this issue, lightweight cryptography has been developed with the objective of designing new ciphers that can be efficiently implemented on these devices [3]. A critical factor in symmetric algorithms is the number of rounds, which directly impacts the computational resources required by the cipher. This parameter is typically established following extensive cryptanalysis to ensure that potential attacks are effectively mitigated. Notably, two prevalent attacks on symmetric ciphers are linear cryptanalysis and differential cryptanalysis [4]. For instance, in the case of AES, the authors determined the number of rounds by considering various cryptanalysis techniques, including linear, differential, truncated differential, and integral cryptanalysis [8]. The number of rounds was set to the maximum number of rounds for which any such attack is applicable, plus an additional security margin. Naturally, by

© The Author(s), under exclusive license to Springer Nature Switzerland AG 2025
W. Meng et al. (Eds.): ADIoT 2024, LNCS 15397, pp. 155–167, 2025.
https://doi.org/10.1007/978-3-031-85593-1_10

exploring various methods to mitigate these attacks, the efficiency of the cipher can be directly improved by reducing the number of rounds.

Moving Target Defense (MTD) is a recently proposed philosophy for improving systems security in general. The motivation behind MTD is that the static nature of some systems, e.g., hardware, well-known IP addresses, etc., gives an attacker time to find and try different ways to counterattack the security measures put in place. MTD tries to mitigate this by reducing the attacker's time window. Although numerous MTD approaches exist for network security, such as IP shuffling, port hopping, and programming language diversity [7], there are relatively few MTD applications in cryptographic systems. One such example is switching between multiple cryptosystems to reduce the success probability of brute force attacks, particularly on resource-constrained devices [6]. However, this approach can be difficult to implement on legacy systems and requires the deployment of numerous cryptosystems to enhance security, which is challenging for resource-constrained devices.

In this paper, we propose Singularization, a novel framework for designing block ciphers based on MTD principles. The core concept of our method is to employ a dynamically generated pseudorandom function (PRF) for each round of a Feistel network. Previously used to improve security applets on legacy SIM cards, Singularization now generalizes to entire families of symmetric cryptographic algorithms [15].

2   Related Works

The concept of incorporating dynamic elements into the structure of a cipher is well-established, though these elements are typically associated with the input data rather than the algorithm itself. Similar Feistel-based methods generally employ a limited number of pseudorandom functions (PRFs), such as 2 or 3, which are combined in a predetermined manner during each execution of the algorithm. Intuitively, we say that a cipher has a static structure if the algorithms remain the same at different runs. Another way of thinking of this is that the cipher can be implemented without any control structures such as IF-THEN statements.

Skipjack is a block cipher designed as an unbalanced Feistel network [14]. The cipher executes distinct operations in even and odd rounds. During even-numbered rounds, it employs a substitution-permutation network referred to as an A-cycle. In odd-numbered rounds, it utilizes a different substitution-permutation network known as a B-cycle. Despite these variations in operations based on the round number, the overall algorithm remains static, meaning that the cipher consistently follows the same sequence of steps in each execution, with the specific steps differing between even and odd rounds.

CAST-256 is another Feistel network that segments the input into multiple blocks [1]. The cipher runs 48 rounds, which are executed into three groups of 16 rounds each. For every group, the cipher performs different operations. As in the case of Skipjack [14], although changes are performed in the steps taken at different rounds, overall, its structure is static, i.e., the algorithm runs the same steps regardless of the key or the input.

Blowfish is a Feistel network featuring a variable key length, the closest idea to our approach [13]. The cipher initializes the S-boxes in a key-dependent manner, causing the S-boxes to change with each execution based on the key. This approach is more dynamic than those used in CAST-256 [1] and Skipjack [14], although it remains static according to our informal definition. While the internal state of the cipher is altered using different tables generated by the key, the operations performed on this data remain consistent. For instance, consider the F-Function used in each round. Although the four S-boxes change according to the key, the round's output is generated solely through XOR operations. Consequently, the algorithm can be implemented without any control structures once the S-boxes are instantiated.

In contrast to previous methods, our framework generates ciphers that dynamically produce the pseudorandom function (PRF) used in each Feistel round based on a secret key. It is important to note that the PRFs for each round differ from an algorithmic perspective; i.e., different keys will result in different encryption algorithms. This approach aims to ensure the secrecy of both the encryption key and the encryption algorithm. Consequently, an attacker must first identify the encryption algorithm (i.e., determine which PRF is used in each round) before attempting an attack. The remainder of the paper is structured as follows: Sect. 3 outlines the general framework of Singularization. Section 4 presents a case study on a reduced 6-round DES, demonstrating how redesigning the cipher within our framework renders differential cryptanalysis as complex as a direct brute-force attack on the key. Finally, Sect. 5 provides conclusions and suggests directions for future research.

3   Cipher Structure

1. $S_0||S_1$ denotes the concatenation between the binary strings S_0 and S_1.
2. $S_0 \oplus S_1$ denotes the result of the bitwise XOR operation between the binary strings S_0 and S_1 of equal lengths.
3. SPLIT (S) denotes the procedure that receives as input a binary sequence S of $2n$ bits and returns two binary strings containing the most and least significant n bits of S:

$$S_L, S_R \leftarrow \text{SPLIT}(S)$$

where $S_L||S_R = S$.

General Structure. A cipher instantiated by Singularization is based on the structure of a Feistel network [10]. Following the MTD philosophy, we sought to increase the complexity of a possible attack by using, for each round, a pseudo-random function chosen uniformly at random from a set of distinct pseudoran-dom functions. Let $F_c : \{0,1\}^{k_c} \times \{0,1\}^n \rightarrow \{0,1\}^n$, $c \in \mathbb{N}$, be a PRF with a key of length k_c bits and a block of length n bits. We consider the following set of N_s PRFs:

$$\mathbf{F} = \{F_1, F_2, \ldots, F_{N_s}\}$$

We denote by $\mathbf{F}_i$, the ith PRF in the set, i.e., F_i. Note that all PRFs in the set $\mathbf{F}$ are different. In theory, two PRFs can be considered distinct if they use different keys. However, our design goes beyond merely changing the key from one round to another; it involves altering the underlying algorithms that implements the PRFs. In this sense, we define two PRFs as distinct as follows:

Definition 1. *Let $F_x : \{0,1\}^{k_x} \times \{0,1\}^n \rightarrow \{0,1\}^n$ and $F_y : \{0,1\}^{k_y} \times \{0,1\}^n \rightarrow \{0,1\}^n$ be two pseudorandom functions with key lengths of k_x and k_y bits respectively. Both functions act over binary strings of length n. We called F_x and F_y ϵ-different, denoted by $F_x \neq_\epsilon F_y$, if:*

$$Pr\left[F_x\left(k,m\right) = F_y\left(k,m\right)\right] < \epsilon \forall \left(k,m\right) \in \{0,1\}^{k_c} \times \{0,1\}^n$$

The definition aims to capture the intuition that two PRFs are considered distinct if their underlying computations differ. We formally define two compu-tations as different if the probability of returning the same outputs for the same inputs is below a certain threshold, denoted by ϵ. At each round i of the cipher, we select a PRF, F_i, uniformly at random from the set $\mathbf{F}$. The input to the cipher is $2n$ bits in length, divided into two parts: the left part and the right part, each consisting of n bits. The left part represents the most significant bits of the input, while the right part represents the least significant bits. Similarly, the output of the cipher is $2n$ bits long and is divided in the same manner as the input. The total number of rounds is denoted by N_r.

The Round Operations. We use two types of keys in a single round:

1. The *round key*, RK_i, represents the key used by the PRF chosen in round i, $1 \leq i \leq N_r$.
2. The *selection key*, SK_i, represents the index of the selected PRF from the set $\mathbf{F}$ in round i, $SK_i \in \{1, 2, \ldots, N_s\}$.

We call the state any intermediate result of the cipher. The state S_i represents the output of round i. The operations performed in a single round are described by ROUNDFUNCTION in Algorithm 1. The ROUNDFUNCTION receives as inputs the current state, S_i, the round key, RK_i and the selection key SK_i based on which the PRF of the round is determined. The inverse of a round function is described by INVERSEROUNDFUNCTION in Algorithm 2.

Algorithm 1. Round operations

1: **function** ROUNDFUNCTION(S_i, RK$_i$, SK$_i$)
2:    $S_{L_i}, S_{R_i} \leftarrow$ SPLIT(S_i)
3:    $S_{L_{i+1}} \leftarrow S_{R_i}$
4:    $S_{R_{i+1}} \leftarrow \mathbf{F}_{SK_i}(RK_i, S_{R_i}) \oplus S_{L_i}$
5:    **return** $S_{L_{i+1}}||S_{R_{i+1}}$
6: **end function**

Algorithm 2. Inverse round operations

1: **function** INVERSEROUNDFUNCTION(S_i, RK$_i$, SK$_i$)
2:    $S_{L_i}, S_{R_i} \leftarrow$ SPLIT(S_i)
3:    $S_{R_{i-1}} \leftarrow S_{L_i}$
4:    $S_{L_{i-1}} \leftarrow \mathbf{F}_{SK_i}(RK_i, S_{L_i}) \oplus S_{R_i}$
5:    **return** $S_{L_{i-1}}||S_{R_{i-1}}$
6: **end function**

Encryption and Decryption. The encryption and decryption procedures of the cipher, ENCRYPT and DECRYPT, are described in Algorithms 3 and 4 respectively. The encryption/decryption algorithm receives the plaintext/ciphertext as input, the set of the round keys, and the set of the selection keys.

Algorithm 3. Encryption

1: **function** ENCRYPT($P, \{RK_1, RK_2, \ldots, RK_{N_r}\}, \{SK_1, SK_2, \ldots, SK_{N_r}\}, N_r$)
2:    $S_1 \leftarrow$ ROUNDFUNCTION($P,$ RK$_1,$ SK$_1$)
3:    **for** i = 2 **to** N$_r$ **do**
4:        $S_i \leftarrow$ ROUNDFUNCTION($S_{i-1},$ RK$_i,$ SK$_i$)
5:        $i \leftarrow i + 1$
6:    **end for**
7:    **return** S_{N_r}
8: **end function**

Design Rationale. We designed this cipher following the moving target defense strategy [7]. There are four main ideas of our design:

1. Our cipher is based on a Feistel network to facilitate theoretical analysis and formal proofs based on previous works [10].
2. At every round, we chose the PRF uniformly at random from a set of PRFs following the idea of changing between multiple cryptosystems proposed in [6]. This will change the cryptographic-related code that is called each round.
3. We set the number of rounds, N_r to be at least four to facilitate the theoretical analysis of our cipher in the context of Luby-Rackoff theorem [11].

A generic cipher designed with Singularization is illustrated in Fig. 1.

Algorithm 4. Decryption

1: **function** DECRYPT($C, \{RK_1, RK_2, \ldots, RK_{N_r}\}, \{SK_1, SK_2, \ldots, SK_{N_r}\}, N_r$)
2:     $S_{N_r} \leftarrow C$
3:     **for** i = $N_r - 1$ **to** 0 **do**
4:         $S_i \leftarrow$ INVERSEROUNDFUNCTION($S_{i+1}, RK_{i+1}, SK_{i+1}$)
5:         $i \leftarrow i - 1$
6:     **end for**
7:     **return** S_0
8: **end function**

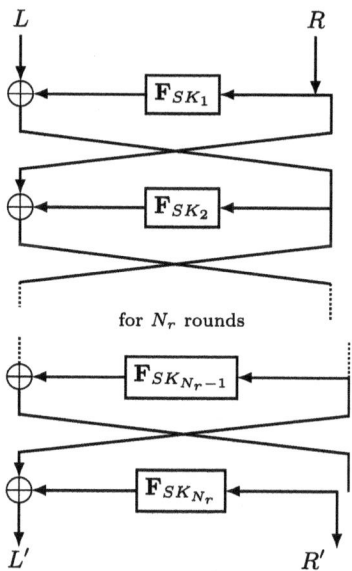

Fig. 1. The cipher structure

4   Case Study

In this section, we present a case study on 6-round DES [12]. We illustrate how this variant of DES is vulnerable to differential cryptanalysis and how the attack is mitigated by redesigning the cipher in the proposed framework [5,9]. We denote by P^i the difference between two inputs of round i, defined as the bitwise XOR of these inputs. Correspondingly, C^i represents the difference between the outputs of round i. Additionally, P_L^i and P_R^i denote the most significant and least significant bits of P^i, respectively, while C_L^i and C_R^i denote the most significant and least significant bits of C^i, respectively. The notation $(a_0, a_1, a_2, a_3)_x$ represents an array of 4 bytes in hexadecimal format.

4.1   Differential Cryptanalysis

Differential cryptanalysis aims to exploit instances where a specific difference between two plaintexts likely results in a particular difference between the corresponding ciphertexts. In an ideally secure cipher, given a plaintext difference $P = P' \oplus P''$, the probability of obtaining the corresponding ciphertext difference $C = C' \oplus C''$ is $\frac{1}{2^n}$, where n represents the block size. Since some of the S-boxes do not produce an equally distributed output, the difference between two inputs of the S-box will not result in an equally distributed difference in the corresponding outputs. A difference distribution table captures how various input differences generate specific output differences. The element on row P and column C counts how many input pairs (P', P'') with $P' \oplus P'' = P$ produces a pair of outputs (C', C'') with $C' \oplus C'' = C$. If an S-box does not have equally distributed outputs, then we say it is vulnerable to differential cryptanalysis. The propagation of a specific input difference through the rounds of the cipher to a particular output difference is referred to as differential characteristics. Assuming there are vulnerable S-boxes in each round, an input difference of P^i in the ith round will, with high probability, produce a specific difference C^i between the round outputs. The sequence $P^1 \to P^2 \to \cdots \to P^{N_r}$ represents the differential characteristic, where N_r is the number of rounds in the cipher. Knowing P^{N_r}, the input difference of the last round enables the attacker to recover part of the last round key. The target partial subkey refers to all bits of the last round key influenced by vulnerable S-boxes. If the attacker has access to a chosen plaintext oracle, the attack proceeds as follows:

1. Determine the differential characteristics of the cipher: $P^1 \to P^2 \to \ldots, \to P^{N_r}$.
2. Generate a pair of plaintexts, (P', P'') at difference P^1, i.e., $P' \oplus P'' = P^1$.
3. Encrypt the plaintexts into the ciphertexts (C', C'').
4. For each ciphertext form the pair (C', C''), determine the inputs of the last round for each possible target partial subkey.
5. If the difference between the inputs of the last round is equal to the difference expected from the differential characteristics, i.e., P^{N_r}, increment a counter for the current target partial subkey.
6. Determine the correct partial subkey as the one with the highest counter and brute-force the rest of the remaining bits of the last round key.

Table 1. S_1 difference distribution table

In	Out			
	0x0	0x4	0x5	0x8
0x03	14	10	6	6
0x0E	0	6	6	6
0x24	12	2	2	14
0x30	0	12	6	8

4.2 DES Structure and Differential Cryptanalysis

The particular structure of 6-round DES is depicted in Fig. 1. DES uses a 56-bit key with a block size of 64 (Fig. 3).

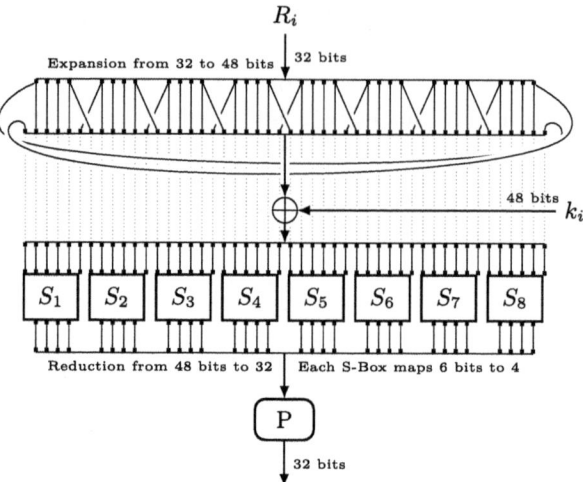

Fig. 2. Round function F

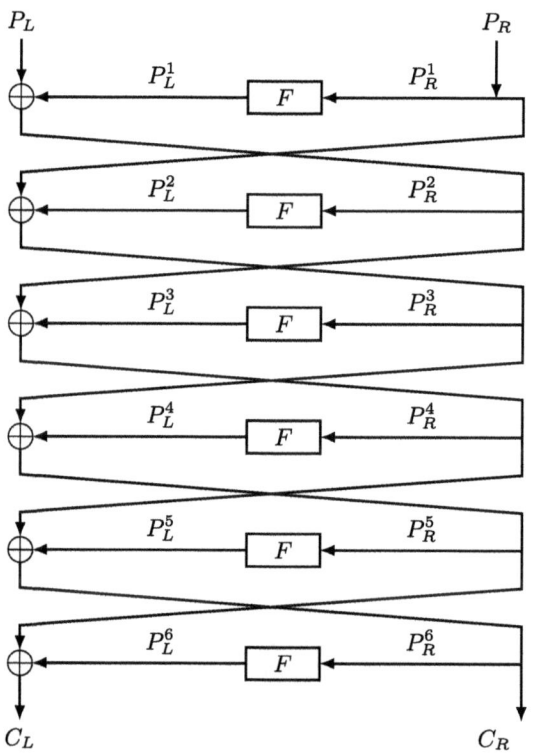

Fig. 3. 6-round DES

The round function, F, is a substitution-permutation network with the following structure:

1. The PRF receives as input a block of 32 bits and a round key of 48 bits.
2. Applies an expansion function on the block input and computes the bitwise XOR between the result and the round key. The expansion function extends the 32 input bits to 64.
3. Runs the result through a set of 8 S-boxes and applies the round permutation.

Each S-box has a 6-bit input and a 4-bit output; therefore, the input and output of the PRF are both 32 bits. The PRF for a round is illustrated in Fig. 2. Given that an S-box has a 6-bit input and a 4-bit output, the difference distribution table will contain 64 rows and 16 columns. Table 1 presents a subset of the difference distribution table for the S_1 substitution. If S_1 produced equally distributed outputs, each element of the table would be equal to 4. However, this is not the case for the S_1 substitution. For example, given an input difference of 0x03, instead of having a probability of $\frac{4}{64}$ for the output 0x0, we observe a probability of $\frac{14}{64}$.

Table 2. First differential characteristics

i	P_L^i	P_R^i	Probability
1	$(40\ 08\ 00\ 00)_x$	$(04\ 08\ 00\ 00)_x$	1
2	$(04\ 00\ 00\ 00)_x$	$(40\ 08\ 00\ 00)_x$	1/4
3	$(00\ 00\ 00\ 00)_x$	$(00\ 00\ 00\ 00)_x$	1
4	$(04\ 00\ 00\ 00)_x$	$(40\ 08\ 00\ 00)_x$	1/4

Table 3. Second differential characteristics

i	P_L^i	P_R^i	Probability
1	$(00\ 20\ 00\ 08)_x$	$(00\ 00\ 04\ 00)_x$	1
2	$(00\ 00\ 04\ 00)_x$	$(00\ 20\ 00\ 08)_x$	1/4
3	$(00\ 00\ 00\ 00)_x$	$(00\ 00\ 00\ 00)_x$	1
4	$(00\ 00\ 04\ 00)_x$	$(00\ 20\ 00\ 08)_x$	1/4

For DES, two differential characteristics for the first three rounds are described in Tables 2 and 3. Consider, for example, the first differential characteristic. In this scenario, the input difference to the fourth round PRF is $(40\ 08\ 00\ 00)_x$. After expansion, five S-boxes (S_2, S_5, S_6, S_7, S_8) receive zero difference inputs and thus return zero difference outputs. The attacker is interested in determining the output difference of the last round PRF (since this

PRF applies the last round key), i.e., P_L^6. Since the attacker knows the output difference of the cipher, C_L, they can compute P_L^6 as

$$P_L^6 = C_L \oplus P_R^5 \tag{1}$$

On the other hand, P_R^5 is computed from the output difference of the fourth round PRF, P_L^4, and the input difference of the third round PRF, P_R^3, which is known from the differential characteristics:

$$P_R^5 = P_L^4 \oplus P_R^3 \tag{2}$$

Since five S-boxes from the fourth round have zero output difference, 20 out 32 bits of P_L^4 are zero. Tracing back the result, the attacker can compute 20 bits of P_L^6 as:

$$P_L^6 = C_L \oplus P_R^3 \tag{3}$$

The remaining 12 bits can be easily brute-forced. At this point, the attacker knows the input of the last round PRF, $P_R^6 = C_R$, and, due to the vulnerable S-boxes, also knows the output of this PRF, P_L^6. Consequently, the attacker can brute-force the corresponding bits from the target partial subkey for each vulnerable S-box. The main idea is that, unlike a brute-force attack on the 30 bits of the target partial subkey, the attacker determines the corresponding bits from the round key independently for each of the five vulnerable S-boxes. This reduces the number of trials from 2^{30} to 5×2^6.

The attacker proceeds similarly for the second differential characteristic, with the vulnerable S-boxes S_1, S_2, S_4, S_5, and S_6. Since three S-boxes are common between the two differential characteristics (S_2, S_5, S_6), the attacker can determine 42 bits of the encryption key using differential analysis and brute-force only 14 bits.

4.3   DES Redesign

Redesigning the cipher using our framework mitigates the differential cryptanalysis attack by making its complexity comparable to that of a direct brute-force attack on the encryption key. One root cause of the attack is the static nature of the PRF used in each round.

Using our framework, we design the unique PRF at each round with the following specifications:

1. The PRF receives as input a block of 32 bits and a master round key of 48 bits.
2. In addition to the round key of 48 bits, a selection key of 8 bits is also generated.
3. Applies the expansion function on the 32-bit input block to get a 48-bit result.
4. Splits the result of the expansion function and the round key into 6-bit words.
5. Combines the wth word from the expansion result with the wth word from the round key in the following key:

(a) If the wth bit from the selection key is 0, the combination result is the bitwise XOR between the word from the expansion result and the word from the round key.

(b) If the wth bit from the selection key is 1, the combination result is the addition modulo 64 between the word from the expansion result and the word from the round key.

6. Runs the result through the permuted set of S-boxes and applies the round permutation.

Within the context of our framework, the set $\mathbf{F}$ comprises 256 PRFs. The PRF $\mathbf{F}_{SK}$ partitions the input and the round key into arrays of words and combines them by performing either bitwise XOR or addition modulo 64 at the word level. The selection key, SK, determines the specific operations to be applied.

Using the redesigned round function with a PRF chosen uniformly at random for each round, the attacker must determine how the result of the expansion is combined with the key at each round. The differential cryptanalysis attack is successful in the initial DES construction because the mixing between the round key and the input preserves the input difference:

$$(P' \oplus K) \oplus (P'' \oplus K) = P' \oplus P'' \tag{4}$$

From another perspective, for the differential cryptanalysis attack to be successful, the attacker must know the input difference for each S-box. The input for an S-box is the result of mixing the round input, which is known to the attacker, with the round key, which is unknown to the attacker. If the mixing is performed using bitwise XOR, then according to (4), the input differences for the S-boxes are the same as the input difference before mixing. This means that even if the attacker does not know the round key, they can still compute the input difference for the S-boxes. After the redesign, although (4) remains valid if bitwise XOR is replaced by modular addition, the attacker must determine for each word whether the input difference is the result of bitwise XOR or modular addition. In the original DES construction, the attacker must perform 5×2^6 trials to determine the bits of the target partial subkey. In the new variant constructed on our framework, the attacker must perform each of the 5×2^6 trials for each possible mixing operation at each round. Since there are 2^8 possible mixing operations per round (with 8 words to be mixed and two possible operations for each pair of words) and the cipher has six rounds, there are 2^{48} possible mixing operations in total. Consequently, the total number of trials becomes $5 \times 2^6 \times 2^{48} = 5 \times 2^{54}$. This number is comparable to the number of trials required for a brute-force attack on the entire DES key, which is 2^{56}.

5   Conclusions and Further Directions of Research

In this paper, we proposed Singularization, a novel framework for constructing block ciphers inspired by the principles of MTD. Our construction utilizes a

Feistel network where the PRF for each round is randomly selected from a set of PRFs. We examined differential cryptanalysis on a reduced 6-round DES. We demonstrated that redesigning the cipher within this framework reduces the attack's effectiveness to that of an almost direct brute-force attack on the encryption key.

An important next step in advancing the Singularization framework is to generalize the formal protection guarantees by investigating various types of cryptanalysis attacks without limiting the analysis to a specific cipher.

Future research directions include exploring the application of this framework to redesign other encryption algorithms. This could result in ciphers with a reduced number of rounds, thereby enhancing the efficiency of encryption algorithms in terms of running time and energy consumption, which is particularly important for resource-constrained devices.

Another avenue for research is to conduct a comprehensive security analysis of the framework using other types of attacks, such as linear cryptanalysis or side-channel attacks.

References

1. Adams, C., Gilchrist, J.: The cast-256 encryption algorithm. RFC 2612, pp. 1–19 (1999). https://dblp.org/rec/journals/rfc/rfc2612
2. Alluhaidan, A.S.D., Prabu, P.: End-to-end encryption in resource-constrained IoT device. IEEE Access 11, 70040–70051 (2023)
3. Bhagat, V., Kumar, S., Gupta, S.K., Chaube, M.K.: Lightweight cryptographic algorithms based on different model architectures: a systematic review and futuristic applications. Concurr. Comput. Pract. Exper. 35(1), e7425 (2023)
4. Biham, E.: Differential Cryptanalysis, pp. 332–336. Springer, Boston (2011)
5. Biham, E., Shamir, A.: Differential cryptanalysis of des-like cryptosystems. J. Cryptol. 4, 3–72 (1991)
6. Casola, V., De Benedictis, A., Albanese, M.: A moving target defense approach for protecting resource-constrained distributed devices. In: 2013 IEEE 14th International Conference on Information Reuse & Integration (IRI), pp. 22–29. IEEE (2013)
7. Cho, J.H., et al.: Toward proactive, adaptive defense: a survey on moving target defense. IEEE Commun. Surv. Tutor. 22(1), 709–745 (2020)
8. Daemen, J., Rijmen, V.: AES proposal: Rijndael (1999)
9. Heys, H.M.: A tutorial on linear and differential cryptanalysis. Cryptologia 26(3), 189–221 (2002)
10. Hoang, V.T., Rogaway, P.: On generalized feistel networks. In: Annual Cryptology Conference, pp. 613–630. Springer (2010)
11. Luby, M., Rackoff, C.: How to construct pseudorandom permutations from pseudorandom functions. SIAM J. Comput. 17(2), 373–386 (1988)
12. Pub, F.: Data encryption standard (des). FIPS PUB, pp. 46–3 (1999)
13. Schneier, B.: Description of a new variable-length key, 64-bit block cipher (blowfish). In: International Workshop on Fast Software Encryption, pp. 191–204. Springer (1993)

14. of Standards, N.I., (NIST), T.: Skipjack and kea algorithm specifications. Technical report (1998). https://csrc.nist.gov/Presentations/1998/Skipjack-and-KEA-Algorithm-Specifications
15. Chrystel, G., Gilles, M.-R., Simona, D., Jean-Philippe, W., Alain, Cuaboz.: Position paper: strengthening applets on legacy SIM cards with singularization, a new moving target defense strategy. In: International Conference on Mobile, Secure, and Programmable Networking, pp. 71–74. Springer (2023)

Author Index

W. Meng et al. (Eds.): ADIoT 2024, LNCS 15397, p. 169, 2025.
https://doi.org/10.1007/978-3-031-85593-1

GPSR Compliance

The European Union's (EU) General Product Safety Regulation (GPSR) is a set of rules that requires consumer products to be safe and our obligations to ensure this.

If you have any concerns about our products, you can contact us on ProductSafety@springernature.com

In case Publisher is established outside the EU, the EU authorized representative is:

Springer Nature Customer Service Center GmbH
Europaplatz 3
69115 Heidelberg, Germany

Batch number: 09409845

Printed by Printforce, the Netherlands